Renewable Energy Sources

NIPA® GENX ELECTRONIC RESOURCES & SOLUTIONS P. LTD.
New Delhi-110 034

About the Authors

Dr. N. L. Panwar is an Associate Professor and Head, Department of Renewable Energy Engineering in the Faculty of Engineering at MPUAT, Udaipur, Rajasthan (India). He has been awarded PhD from the Centre for Energy Studies, Indian Institute of Technology Delhi, India. His areas of interest cover energy and exergy analyses of thermal systems and energy efficiency and management. He has contributed more than 150 papers in international journals and 15 books on renewable energy aspect. He has been a recipient of 'Prakritik Urja Puraskar' from Ministry of New and Renewable Energy, Government of India for his outstanding book on alternative energy resources. He has also been awarded by "Shrimati Vijay-Usha Sodha Research Award" from the Indian Institute of Technology, Delhi in 2014, and Rajasthan Energy Conservation Award 2018 from the Government of Rajasthan. ***According to Stanford University USA report undersigned, based on citations metrics, is amongst the top 2% of scientists across all disciplines of research in the world.***

Dr. Sunil L Narnaware is an Assistant Professor and In-charge Head, Department of Agriculture Engineering in the Faculty of Engineering and Technology, at Datta Meghe Institute of Higher Education and Research (Deemed to be University), Sawangi (M), Wardha, Maharashtra. (India). He pursued PhD at Maharana Pratap University of Agriculture and Technology, Udaipur. (Raj.). He has more than 11 years of experience in the field of research, teaching as well as in renewable energy industry. He has published more than 12 papers in national and international journals.

Er. Swati Narnaware is working as an Assistant Professor in the Faculty of Engineering and Technology, at Datta Meghe Institute of Higher Education and Research (Deemed to be University), Sawangi (M), Wardha, Maharashtra. She has completed her M.Tech. from Junagarh Agricultural University, Junagarh, Gujarat. She has more than 8 years of experience in Research and Development in solar thermal system design and testing. As an Incubation manager at Dr. PDKV, Akola, she handled national and international level training, seminars, and workshops for students and faculties of agricultural and allied sciences. She had taken a lead role in the development of training modules for students and faculties. She was involved in the development of an e-module for undergraduate courses in Agricultural Engineering.

Renewable Energy Sources

N.L. Panwar
Associate Professor & Head
Department of Renewable Energy Engineering
Mahararan Pratap University of Agriculture and Technology
Udaipur (Rajasthan) India

Sunil L Narnaware
Assistant Professor
Department of Agriculture Engineering
Faculty of Engineering and Technology
Datta Meghe Institute of Higher Education and Research (Deemed to be University)
Sawangi (M), Wardha, Maharashtra. (India)

Swati Narnaware
Assistant Professor
Department of Agriculture Engineering
Faculty of Engineering and Technology
Datta Meghe Institute of Higher Education and Research (Deemed to be University)
Sawangi (M), Wardha, Maharashtra. (India)

NIPA® GENX ELECTRONIC RESOURCES & SOLUTIONS P. LTD.
New Delhi-110 034

NIPA® GENX ELECTRONIC RESOURCES & SOLUTIONS P. LTD.

101,103, Vikas Surya Plaza, CU Block
L.S.C. Market, Pitam Pura, New Delhi-110 034
Ph : +91 11 27341616, 27341717, 27341718
E-mail: newindiapublishingagency@gmail.com
Website: www.nipabooks.com

For customer assistance, please contact

Phone: + 91-11-27 34 17 17
Fax: + 91-11-27 34 16 16
E-Mail: feedbacks@nipabooks.com

ISBN: 978-81-19072-11-8

Composed and Designed by NIPA®.

Preface

There is no doubt that energy plays a crucial role in the development and progress of the national economy. In response to rapid industrialization, infrastructure expansion, population growth, as well as the expansion of the economy, there is a need for more sources of energy. Even so, conventional energy reserves have a finite lifespan and will only become available in a certain number of years with steadily increasing prices in the future. Additionally, the use of fossil fuels contributes to the development of environmental problems, particularly the global warming problem. Our current energy market presents us with an opportunity to reduce our dependence on fossil fuels by expanding and diversifying our energy supply mix. In addition, we can shift the development process towards sustainability as well as environmental and social responsibility in light of the challenges facing us today. These new energy sources offer us a great opportunity to reduce our dependency on fossil fuels. In order to achieve this goal, it is essential to have a basic understanding of renewable energy technology that can be used to achieve it. In this book, there is a systematic approach to presenting a systematized overview of renewable energy resources and devices.

It is intended for engineering and science students from a wide range of backgrounds who are interested in studying the engineering aspects of renewable energy in depth. A total of thirteen chapters are provided in this book, each dealing with a different aspect of renewable energy engineering. This book provides a basic understanding of renewable energy sources through the information presented. This book can be used as a textbook by undergraduate and postgraduate students, as well as professionals who are involved in the renewable energy sector.

Udaipur Wardha **Authors**

Contents

Preface v

1. Introduction to Energy 1

1.1. Introduction 1
1.2. World Energy Scenario 1
1.3. Energy Sources 3
1.4. Commercial Energy Sources 4
1.5. Role of Renewable Energy 6
1.6. Greenhouse Effect 7
1.7. Global Warming 8
1.8. Conservation of Fuel 8

2. Energy in India 9

2.1. Energy Consumption Pattern and Energy Resources in India 9
2.2. Non-Commercial Energy 9
2.3. Commercial Energy Resources 9
2.4. Consumption of Coal and Lignite 13
2.5. Consumption of Crude oil and natural gas 13
2.6. Consumption of Electricity 13

3. Solar Energy 15

3.1. Introduction 15
3.2. Solar Radiation 15
3.3. Solar Radiation at the Earth's Surface 16
3.4. Atmospheric Effects 17
3.5. Solar Constant 18
3.6. Air Mass 18
3.7. Absorption in the Atmosphere 19
3.8. Type of Radiations 19
3.9. Solar Energy Technologies 20

3.10. Solar Thermal Application 21
3.11. Types of Solar Collectors 21
3.12. Transport fluid options 23
3.13. Collector Construction 24
3.14. Energy Balance in Flat-Plate Collectors 25
3.15. Flat Plate Collector Performance 25
3.16. Concentrating Solar Collectors 26
3.17. Solar Thermal Power Systems 31
3.18. Solar Photovoltaic 36
3.19. Cells, Modules, Panels and Arrays 38
3.20. Advantages and Disadvantages of PV System 39
3.21. Types of Solar Cell 39
3.22. I-V Characteristic 40
3.23. Series and Parallel Connection of Solar Cell 42
3.24. Electrical Characteristics of a Photovoltaic Array 42
3.25. Efficiency of Solar Cell 43
3.26. Solar PV Plant 44
3.27. Solar PV System Sizing 46

4. Wind Energy 51
4.1. Introduction 51
4.2. The nature of Wind 51
4.3. Potential of Wind Energy in India 53
4.4. Relationship Between Wind Speed and Height 53
4.5. The Power in Wind 54
4.6. Power Coefficient 55
4.7. Cut-in and Cut-out Speed 56
4.8. Maximum Power in Wind 56
4.9. Forces on the Blades and Thrust on Turbines 58
4.10. Types of Wind Turbines 59
4.11. Horizontal Axis Wind Turbines (HAWT) 60
4.12. Components of Wind Mill 61
4.13. Vertical Axis Wind Turbines(VAWT) 63
4.14. VAWT Subtypes 65

4.15. Wind Farm 67
4.16. Site Selection for Wind Farm 69

5. Biomass Energy 77
5.1. Introduction 77
5.2. Biomass to Energy Conversion 78
5.3. Biomass Combustion System 79
5.4. Biomass Combustion Power Plant 82
5.5. Benefits and Drawbacks of Biomass 83
5.6. Scope of Biomass Based Power Generation in India 84
5.7. Technology Used for Biomass Based Power Generation 84
5.8. Biomass Gasification System 85
5.9. Why Choose Gasification? 87
5.10. Types of Gasifiers 87
5.11. Factor Affecting Gasification Process 92
5.12. Advantages and Disadvantages of Gasification 92
5.13. Capacities of Gasifier Based System 93
5.14. Scales of Biomass Power Generation 93
5.15. Biomass Pyrolysis 94
5.16. Economic and Environmental Advantages of Pyrolysis 95
5.17. Utilization of Pyrolysis Products 96
5.18. Biomass Pyrolysis Process Considerations: 96
5.19. Bagasse based cogeneration 98
5.20. Efficiency of Biomass-based Power Generation 99

6. Biomass Pelletization and Baling 103
6.1. Introduction 103
6.2. Pelleting Process 103
6.3. Calorific Value of Different Pellets 105
6.4. Factors Affecting Pellet Quality 106
6.5. Government Initiatives 107
6.6. Balers for Agricultural Crop Residues Management 108
6.7. Baler Machine 108
6.8. Functioning of a Baler 108
6.9. Types of Baler Machines in Agriculture 108

7. Biodiesel Production 111
7.1. Introduction 111
7.2. Raw Materials for Biodiesel Production 111
7.3. Biodiesel Production Process 112
7.4. Transesterification for Biodiesel Conversion 112
7.5. Characteristics of Biodiesel 114
7.6. Prospective Feedstock in India 116
7.7. Economic Feasibility 116
7.8. Biodiesel's advantages 117
7.9. Disadvantages of Biodiesel 117

8. Biogas Technology 121
8.1. Introduction 121
8.2. Anaerobic Digestion 121
8.3. Benefits of Anaerobic Digestion 124
8.4. Components of Biogas Plant 124
8.5. Types of Biogas Plants 124
8.6. Biogas Application 128
8.7. Power Generation from Biogas 129
8.8. Commercial Application of Biogas Plant 131

9. Fuel Cell 137
9.1. Introduction 137
9.2. Application of Hydrogen 137
9.3. Advantages and Limitation of Hydrogen Fuel 138
9.4. Fuel Cell Technology 138
9.5. Fuel Cell 139
9.6. Classification of Fuel Cells 140
9.8. Application of Fuel Cell 146
9.9. Fuel Cell Thermodynamics 147

10. Micro Hydropower 149
10.1. Introduction 149
10.2. Classification of Small Hydro Project 149
10.3. Determination of the Power Output 151
10.4. Use of Micro-hydro Power Plant 151

10.5. Working of Micro-hydropower System 152
10.6. Types of Turbines Used 153

11. Ocean Thermal Energy Conversion 159
11.1. Introduction 159
11.2. Closed cycle OTEC 160
11.3. Open-Cycle OTEC 160
11.4. Hybrid OTEC 161

12. Magneto Hydrodynamic 165
12.1. Introduction 165
12.2. MHD Generator 165
12.3. Principle of MHD Generation 166
12.4. Components of the MHD 166
12.5. The Faraday Current 167
12.6. Hall Effect Current 167
12.7. Power Calculation of MHD 167
12.8. Types of MHD Cycles 168
12.9. Advantages of MHD System 170
12.10. Disadvantages of MHD System 171

13. Energy Conservation in Agriculture 173
13.1. Introduction 173
13.2. Farm Energy Management Approach 174
13.3. Energy Conservation measures in the Agricultural Sector 175
13.4. Energy Conservation by Use of New and Renewable Sources of Energy 187
13.5. Some Tips for Energy Conservation 191

References 197

Index 199

1

Introduction to Energy

1.1 Introduction

The life on the earth is driven by the energy available in the various forms. Energy is the most crucial entity in development of the world. The sufficient and sustainable supply of energy is deciding factor in the global energy market. Solar energy from the sun is the basic source of energy on the earth which further manifest into the other energy forms. However, the fossil fuel energy is widely used energy on the larger scale on commercial basis due to its availability and technological advancement in handling these fossil fuels.

1.2 World Energy Scenario

The growth in population and energy consuming sectors (Figure 1.1), the economy has seen the tremendous increased in energy demand. If we look at the regional scenario about the energy demand, the demand comes from the fast growing developing countries which is led by China and India. Industrial sector including manufacturing, construction, refining, mining, and agriculture is the largest energy consuming sectors which is more than 50%. It has been estimated that world industrial sector energy use increases more than 30% between 2018 and 2050 due to increase in consumption of goods and will reach to 315 Quadrillion British thermal unit (Btu) which can be seen in Figure 1.2 and Figure 1.3.

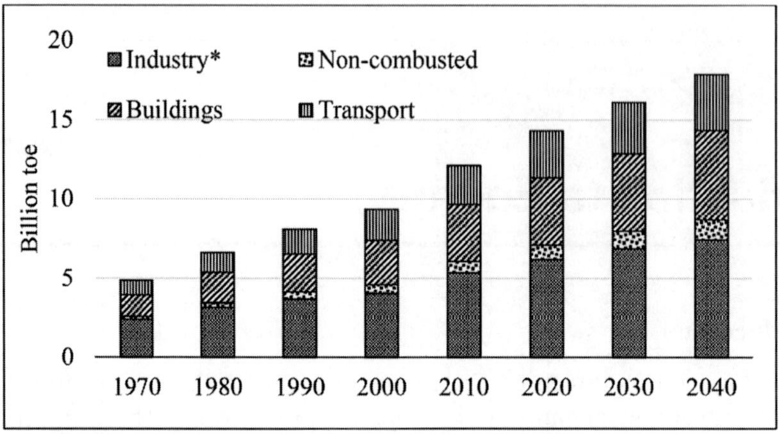

Fig. 1.1: Primary energy demand by end-use sector

*Industry excludes non-combusted use of fuels

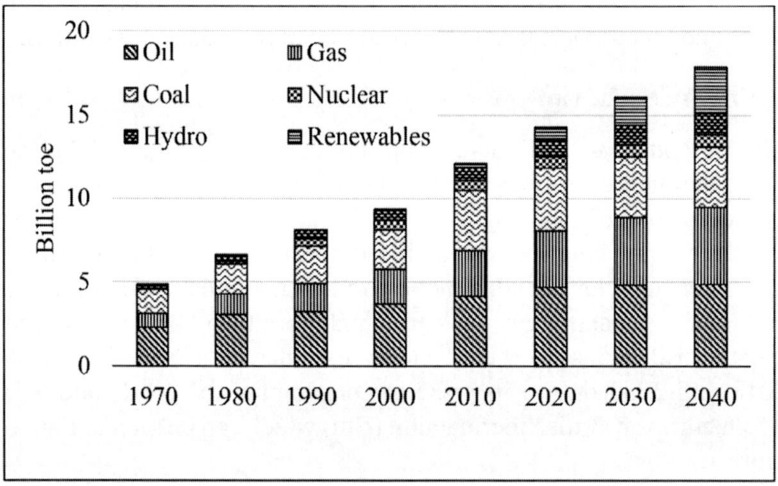

Fig. 1.2: Primary energy demand fuel wise

Over the years, it has been seen the renewable energy is the fastest growing energy sector which is almost accounts for half of the increase in energy. Renewable energy has witnessed an increase of 30% in share of the global power sector where coal declines significantly and projected to surpass by renewable as primary energy source in global power sector. The major contributor in renewable energy are solar and wind power as both are witnessing rapid growth due to decreasing cost.

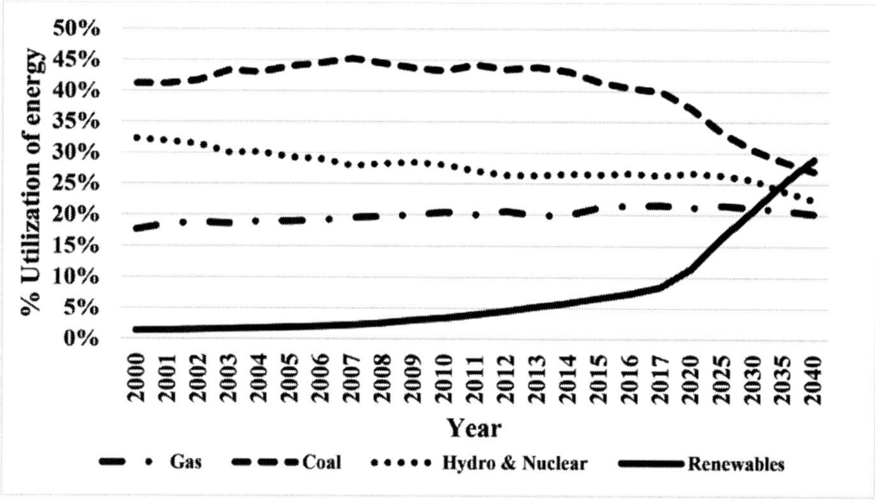

Fig. 1.3: Trend in utilization of different energy sources

1.3 Energy Sources

Energy sources are generally classified broadly in two categories: Conventional energy sources and Non-conventional energy sources. Conventional energy sources are fossil fuel occurring in the interior of the earth available in limited quantity. These energy sources have been formed by decaying matter over hundreds of years and are being used since long time but once depleted they can be generated again. The continuous exploitation has decreased the shortage to a greater extent. Coal, petroleum oil, natural gas are examples

A. Conventional

B. Non-conventional

A. *Conventional Energy Sources*

(i) Commercial

a. Coal

b. Petroleum

c. Electricity

(ii) Non commercial

a. Firewood

b. Straw

c. Dung (dried)

B. ***Non-conventional Energy Sources***

a. Solar

b. Wind

c. Bioenergy

d. Geothermal energy

e. Ocean energy

f. Tidal energy

Non-conventional sources of energy are non-fossilized energy sources which continuously replenished because of the natural process. These energy sources are also called as renewable energy sources or alternate energy sources.

1.4 Commercial Energy Sources

1.4.1 Fossil Fuels

Coal

Coal forms an important fossil fuel and contribute major share of energy in world energy consumption. It is a combustible solid carbon rich material received as a sedimentary rock. It mainly composed of carbon and hydrocarbon and is brown black in color. The formation of coal took millions of years as a result of huge pressure and heat of layers of dirt and rock covered over the plants. So coal contains the energy from the plants that lived hundreds of the millions of years ago in swampy forests. Coal is considered as non-renewable energy source and is the largest resources used for production of electricity. Coal is available in abundant quantity and is used as a source of energy for production of electrical power using steam generation. The chemicals (dyes, oils, waxes, pharmaceuticals and pesticide) can be derived from the coal.

The classification or types of coal is determined based on the carbon content and volatile matter content on percent basis and ranked according to the changes occurred over the time. According to the Hilt's law, deeper the coal seam, makes the higher rank coal as the high pressure and temperature is experienced by the material to transform into carbon. Anthracite is regarded as highest grade coal which has about 86 to 98% carbon and 3 to 8% volatile matter. Bituminous coal contains about 70-86% carbon and 31 to 46% volatile matter; sub-bituminous coal has 70-76% carbon and 53-42% volatile matter. Lignite is low grade of coal which has 25-35% carbon and high content of volatile matters about 53 to 63% along with high percent of moisture content.

Peat is partially consisting of partially decomposed vegetation with little carbonization process and has less carbon content and in technical terms not considered as coal. The presence of high volatiles and other gases like methane and mercury makes it hazardous to environment when used for burning. Coal is the biggest player in electricity generation worldwide which almost has share of 38% of fuelling the global electricity or even more in some countries.

Oil

Many countries have oil reserves producing less or more than others. The countries that produces highest oil are Saudi Arabia, Russia, the United States, Iran and China. Crude oil and liquid produced from fossil fuels which later undergoes the process of refining to give the petroleum products. Petroleum product is a raw material for many product such as plastics, polyurethane, solvents and hundreds of other intermediate and end user goods. In 2017, the total world consumption of petroleum was 98.8 million barrel per day. United States is the largest consumer of petroleum contributing about 20.2% of the total world share followed by China (13.7%); India (4.4%); Japan (4.0%) and Russia (3.7%).

Natural gas

Natural gas is a mixture of gases rich n hydrocarbons. Methane forms the major component of the natural gas. The reserves of the natural gas are found near the other fossil fuel like coal and crude oil reserves deep inside the earth. Natural gas is processed to get the cleaner fuel and then further used for applications. During the processing several by-products are produced which includes like propane, ethane, butane, carbon dioxide and nitrogen which have again many uses. It is mainly used for generation of electricity and heat. Compresses natural gas (CNG) is used in transportation vehicles. Boilers and air conditioner are also uses natural gas. Ammonia is used for making fertilizers.

Hydroelectric power

Hydroelectric power or hydroelectricity is a form of energy harnessed from the moving water to generate electricity. Hydroelectric power necessarily has the huge water reservoir of water. The potential energy of water is used to convert kinetic energy by making flow through gate outlets. The kinetic energy is used to rotate the water turbines to generate electricity. Hydroelectric energy is the most commonly-used renewable source of electricity. China is the largest producer of hydroelectricity. Other top producers of hydropower around the world include the United States, Brazil, Canada, India, and Russia. Hydroelectric power plant is generally located on the rivers by constructing dams to raise the level of the water to create the head required for operation of turbines. Hydropower is the major source of electricity production in the world accounting about 16% of the total electricity production worldwide. More than 150 countries produce hydroelectricity at some capacity though China, Canada, Brazil and the United States combine have the about 50% of the total power produced. China is leading country in hydroelectricity production with total capacity of 341.1 GW in 2017.

Nuclear energy

Nuclear power plants use nuclear fission process to generate the electricity. Nuclear reactors are used to generate the heat which is covered into the steam to drive the steam turbine to produce electric power. The fuel used in nuclear reactor is mostly uranium however thorium also has potential to use in nuclear plant but currently not used. Nuclear power capacity in increasing steadily. Presently 440 nuclear power reactors are in operation in 32 countries with combined capacity of 400 GWe which forms about 10% of the total electrity produced on global basis. About 50 reactors are under construction of which mostly in the Asian region. As per World Energy Outlook 2020 report, it has been projected that the nuclear capacity growth of over 15% will be seen during 2019 to 2040 reaching about 480 GWe. This increased growth majorly seen Asian countries particularly China and India. Nuclear power is being used since 1960s which was boosted in 1970s to 90s. But in 2011, the Fukushima tsunami in Japan turmoil the growth of nuclear power as safety is the major concern. In 2019, the global nuclear power generation reached to 2586 terawatt-hours.

1.5 Role of Renewable Energy

EIA has forecasted an increase of about 50% in energy uses by 2050 due to strong growth in various energy consuming sectors. The various factors responsible for the boost in energy utilization are increased per capita income, the focus on urbanization and access to electricity in urban and rural areas. Renewable energy or alternate energy are important in a view that it can replace the use of fossil fuel to the certain level. The increased concern about rapid depletion of fossil fuel, the environment pollution and sustainable energy supply in the future, renewable energy has gained an obvious importance in past few decades. Climate change and global warming are the burning issues which make world to think about uncontrolled exploitation and burning of the fossil fuel. The most vital resources land, energy and water have been exploited to the extent that it has affected adversely in the form of climate change. The activities related to the economic development have increased the emission of carbon dioxide and other greenhouse gases which drive up the average temperature of the earth and resulting in extreme weather conditions. This has been attributed to the greenhouse effect.

1.6 Greenhouse Effect

Greenhouse effect refers to the warming when the presence of some gases traps the heat. These gases allow the sunlight to enter into the earth atmosphere but hold back the some radiations and heat. The emission of greenhouse gases is responsible to create greenhouse effect on earth atmosphere. The energy that is received from the sun or solar radiation get trapped in the earth's atmosphere. Some of the solar radiations are radiated back to the space from the surface of the earth but get absorbed by the greenhouse gases and re-emits back to the earth in all direction. This causes heating up of the lower atmosphere and surface of the planet earth.

The principle of greenhouse effect (Figure 1.4) generally observed when short wave length light passes through medium like glass or atmosphere which gets absorbed by the surface or object whereas long wave length infrared radiations passes through the medium but could not escape or pass through the medium after re-radiation form the objects. This trapping of longer wave length increases the temperature inside the medium. In case of earth atmosphere, greenhouse effect is heating of the planet's surface as greenhouse gases such as methane, carbon dioxide and water vapour trap the outgoing infrared or thermal radiation and maintains the temperature suitable for existence of life on earth. However, the increase in greenhouse gases emissions from the burning of the fossil fuels increases concentration of greenhouse gases which traps the more infrared radiations and results in increase in overall average temperature of the earth causing global warming.

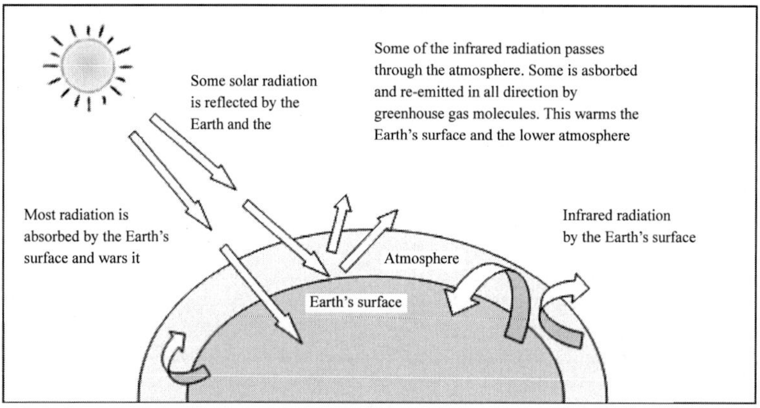

Fig. 1.4: Greenhouse effect

1.7 Global Warming

Global warming is long term warming of planet earth which is increasing the overall temperature. The earth is getting hotter due to increased burning of fossil fuel causing the carbon pollution. According to NASA's Goddard Institute for Space Studies (GISS), the average global temperature on Earth has increased by little more than 1o C since 1880. Two third warming has occurred since 1975, at the rate of roughly 0.15 to 0.20 °C per decade.

During the 150 years of industrial age, the concentration of carbon dioxide in the atmosphere has increased by 31 percent. Global warming has many consequences like change in seasonal natural events, extreme weather, vanishing of polar ice, increase in sea level, coastal floods. Global warming has resulted in an another issue called as climate change which is change in weather pattern and cropping seasons around the world.

Renewable energy sources emit little or no emissions that causes global warming. No combustion cycle is involved in renewable energy which makes them clean and environment friendly. Renewable energy has becoming a promising tool to fight against global warming and climate change.

1.8 Conservation of Fuel

Switching to alternative sources such as solar energy in the forms of solar heater, solar cookers, use of natural gas and wind energy are some ways to reduce their usage. Renewable energy sources, such as wind and solar, emit little to no greenhouse gases, are readily available and in most cases cheaper than coal, oil or gas. Using renewable energy can help to reduce energy imports and reduce fossil fuel use.

2

Energy in India

2.1 Energy Consumption Pattern and Energy Resources in India

Energy is vital commodity in economic development of the world and any country. India also experienced a greater shift from non-commercial to commercial energy use after independence over the period. Industrialization and increased population has increased the consumption of commercial energy sources. The mechanization in agriculture and access to clean energy for household application also increased the consumption of the energy. however, the per capita energy consumption remains lower compared to the developed countries. In broader view, the energy used can be categorised in Non-commercial and Commercial energy sources that are used in India.

2.2 Non-Commercial Energy

Non-commercial energy sources mainly dominate in rural areas and to some extent in urban area as well. Fuelwood forms the major energy source basically used for cooking and heating purposes. Cow dung and crop residues are also used to fulfil the need of energy to some extent. For agriculture operations, draught power or animal power was primary energy source which is replaced to a greater extent by the tractor and other machinery.

2.3 Commercial Energy Resources

Coal and lignite

Coal is most abundant energy resource in India although it has poor quality with high ash content and calorific value. Coal is major commercial energy source used in India and has about 0.8 percent of the total geographical reserves and 5.7 percent of the coal reserve in the world. Coal deposits are mainly confined to eastern and south central part of the country. The states of Jharkhand (25.88%), Odisha (24.76 %), Chhattisgarh, West Bengal, Madhya Pradesh, Telangana and Maharashtra accounts for 98.09% of total coal reserve of coal in country accessed in 2019. In 2019, the total reserve coal in country has been estimated to 326.49 billion tonnes whereas total estimated reserve of lignite were 45.76 million tonnes.

India is world's second largest producer of coal after Peoples Republic of China. The share of coal in both the energy mix and the power mix has been increasing since 1970s. in 2017, coal provided 44% of the total primary energy supply and 74% electricity generation.

Oil and Natural Gas

India has about 0.04 percent of the world's proven reserve of hydrocarbons. As per the estimates, crude oil reserve is about 618.95 million tonnes (MT) within the geographical expansion in Western Offshore (38%) followed by Assam (25.6) whereas the maximum reserve of Natural gas in the Offshore (41%) by the Western offshore (23.4%).

Hydro Power

Government of India, in 1998 announced policy on hydro power development to develop the hydropower in the country. India has immense potential in hydroelectric power generation of about 148700 MW capacity as shown in Table 2.1. This put India at fifth rank in exploitation of hydro potential on global scenario. In addition, 56 number of pumped storage project with cumulative capacity of 94000 MW have been under consideration. Small capacity hydro power projects which comes under small, mini and micro capacity with an estimated capacity of 6782 MW is also scheduled to execution.

Table 2.1: Estimated capacity of hydropower in India

Basin /rivers	**Estimated Installation capacity (MW)**
Indus basin	33832
Ganga basin	20711
Central Indian river system	4152
Eastern flowing rivers of southern India	9430
Eastern flowing rivers of southern India	14511
Brahmaputra basin	66065
Total	148701

Hydropower projects of 25 MW or below capacity are classified into small hydro power which again classified into micro (100 kW), mini (101 -2 MW) and small hydro (2-25 MW). Presently, the hydropower plant upto 3 MW and below comes under the jurisdiction of Ministry of New and Renewable Energy (MNRE). Development of small hydro projects in hilly region has been focused with a targeted capacity of 100MW canal based small hydro power projects through private secor participation. An estimated potential of 21135.37 MW from small and mini hydel project at 7135 sites has been accessed by Alternate Hydro Energy Centre (AHEC) of IIT Roorkee in its Small Hydro Database of July 2016. Hilly states like Arunachal Pradesh, Himachal Pradesh, Jammu Kashmir and Uttarakhand constitute about half of the potential.

Nuclear Resources

The efforts of Dr. Homi Bhabha in early 1944 paved the way for the nuclear energy programme in India. Sir Dorabji Tata charitable trust funded the The Tata Institute of Fundamental Research (TIFR) in 1945 for atomic studies in India. The Department of Atomic Energy under Prime Minister was set up in 1954 to administer the programme of atomic energy. the first nuclear power plant in India was set up at Tarapur, Maharshtra consisting of two boiling water reactors (BWRs) on turnkey basis by GE, USA which was commissioned in October, 1969. In parallel, the work on construction of PHWRs was also commenced. Apart from the first two BWRs at Tarapur which are in operation since 1969, eighteen PHWRs with two reactors at each of the four locations Kalpakkam (MAPS), Narora (NAPS), Kakrapar (KAPS), and Tarapur (TAPS-3&4), four at Kaiga (KGS) and six reactors at Rawatbhata (RAPS) are now in operation. Of these, TAPS- 3&4 are of 540 MWe unit size while the others are in the unit size range of about 200-220 MWe (gross). In addition, one 1000 MWe VVER type PWR (Kudankulam-1) is also operational. The total gross nuclear power capacity in operation is now 5780 MWe. Construction work for one more unit of 1000 MW (e) VVER (Kudankulam-2) is in progress in co-operation with Russian Federation. Several advanced safety features have been provided in these reactors. Construction work for setting up of two units of 700 MWe PHWRs at Rawatbhata (RAPP-7&8) and two units of 700 MWe PHWRs at Kakrapar (KAPP-7&8) are also in progress. The work on the second stage of the nuclear power programme is in progress at the Indira Gandhi Centre for Atomic Research (IGCAR). The Fast Breeder Test Reactor (FBTR) 40 MWth at Kalpakkam is in operation. Its unique carbide fuel has achieved a burn-up of 155,000 MWD/Tonne. Construction of the first 500 MWe Prototype Fast Breeder Reactor (PFBR) is in advanced stage of construction and system wise commissioning trials commenced.

Towards building up thorium-based reactors, the strides taken by DAE include setting up of 30 kWth neutron source reactor KAMINI at Kalpakkam. The reactor has been in operation since 1997. Kamini uses uranium233-based fuel derived from irradiated thorium. A detailed design report for setting up the Advanced Heavy Water Reactor (AHWR) of 300 MWe capacity has already been prepared by BARC. This is a vertical pressure tube reactor design utilising heavy water moderator, boiling light water coolant, thorium-plutonium based fuel and incorporating passive safety systems. It derives about two-third of its power from thorium and DAE/BARC. It will be essentially a technology demonstration project for utilising thorium for electricity generation and the site has been selected and is expected to start construction in a few year time.

Total installed capacity of the power for various sector and for different sources is given in Table 2.2 and 2.3 respectively. Figure 2.1 shows a share of different energy based power generation. Power generation from fossil fuel is still dominant in India.

Table 2.2: Total installed capacity (As on 31.10.2020)

Sector	MW	% of Total
Central Sector	93,927	25.2%
State Sector	103,617	27.8%
Private Sector	175,891	47.1%
Total	3,73,436	

Table 2.3: Power generation installed capacity

Fuel	MW
Total Thermal	2,31,321
Coal	1,99,595
Lignite	6,260
Gas	24,957
Diesel	510
Hydro (Renewable)	45,699
Nuclear	6,780
RES* (MNRE)	89,636
Total	373,436

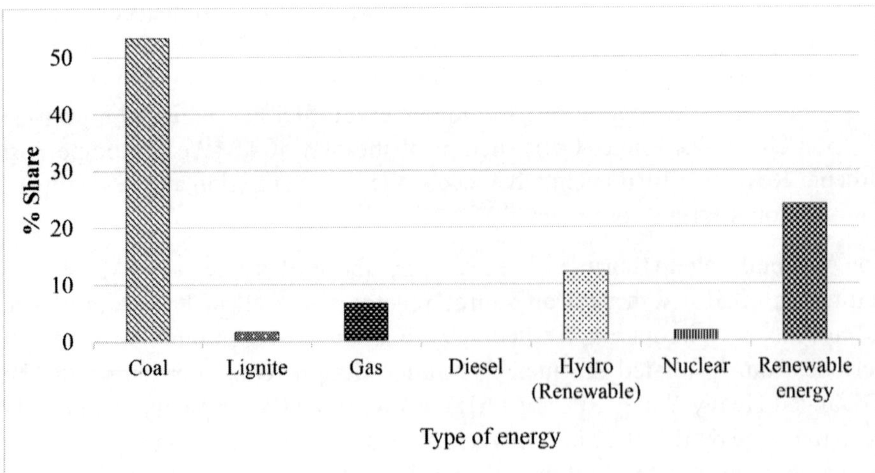

Fig. 2.1: Energy share in consumption on national level (%)

2.4 Consumption of Coal and Lignite

The total consumption of coal by industry from 549.57 MT in 2008-09 to 896.34 MT in 2017-18 which is compound annual growth rate of 5.01 %. Consumption of lignite increased to 45.82 MT in 2017-18 from 31.85 MT in 2008-09. Lignite consumption for electricity production was highest accounting for 83.7% of total lignite consumption. Maximum coal consumption of raw coal was in electricity generation (576.19 MT) followed by steel and washery industries (58.50 MT), sponge iron industry (8.51 MT) and cement industry (7.70 MT)

2.5 Consumption of Crude oil and Natural Gas

The consumption of crude oil was increased from 160.77 MMT during 2008 to 251.53 MMT during 2017-18. The maximum use of natural gas is in fertilizer industry (27.78%) followed by power generation (22.77%) and domestic fuel for transportation sector (16.25%). The major use of natural gas is in energy sector (60.68%) and non-energy (39.32%). Consumption of petroleum products like high speed diesel oil, petrol, pet coke, LPG and Naphtha also used for different sectors as shown in Figure 2.2.

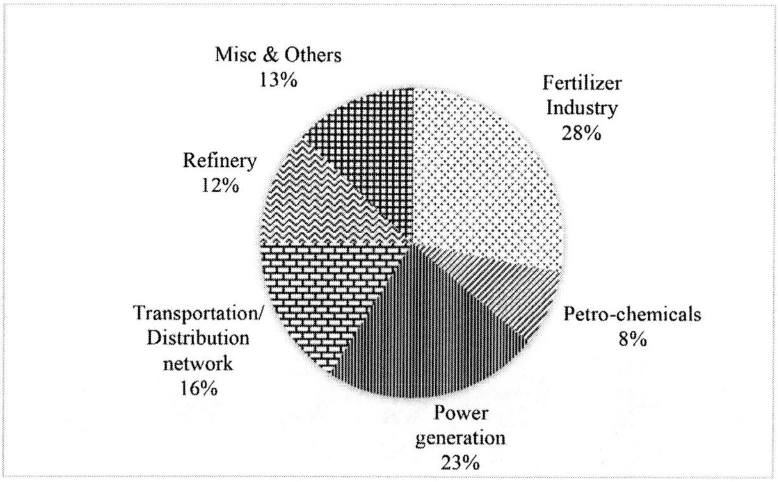

Fig. 2.2: Industry wise off-take of Natural Gas in India during 2017-18

2.6 Consumption of Electricity

Electricity consumption was 1130244 GWh in 2017-18 and seen continuous increase since 2008-09. Sector wise electricity consumption in year 2017-18 is shown in Figure 2.3 which showed that large share of electricity is consumed by industry, domestic and agriculture sector.

Per-capita Energy Consumption (PEC) increased from 19,599 MJ in 2011-12 to 23,355 MJ in 2017-18 (Figure 2.4). Energy is complex issue in India. During 2004-005, about 22% of the urban households depend on firewood, another 22% on kerosene and about 44% on LPG.

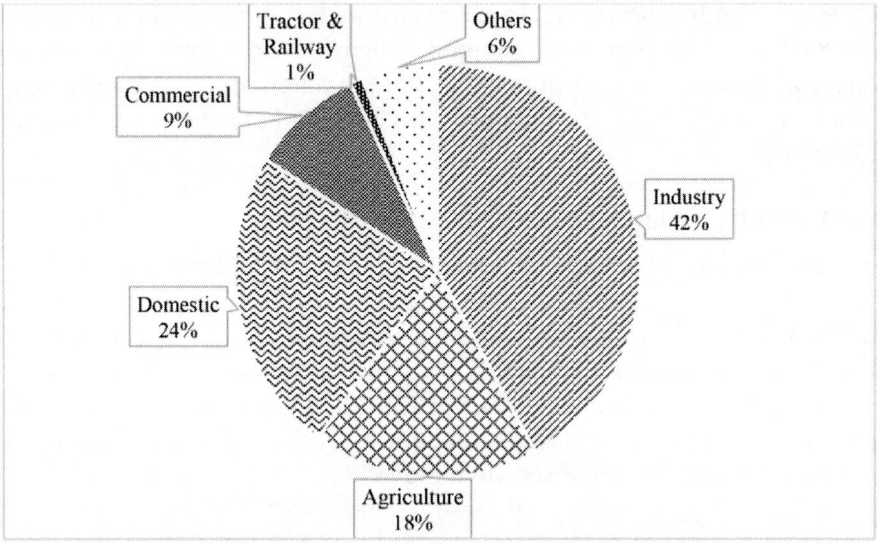

Fig. 2.3: Consumption of electricity by sectors in India during 2017-18

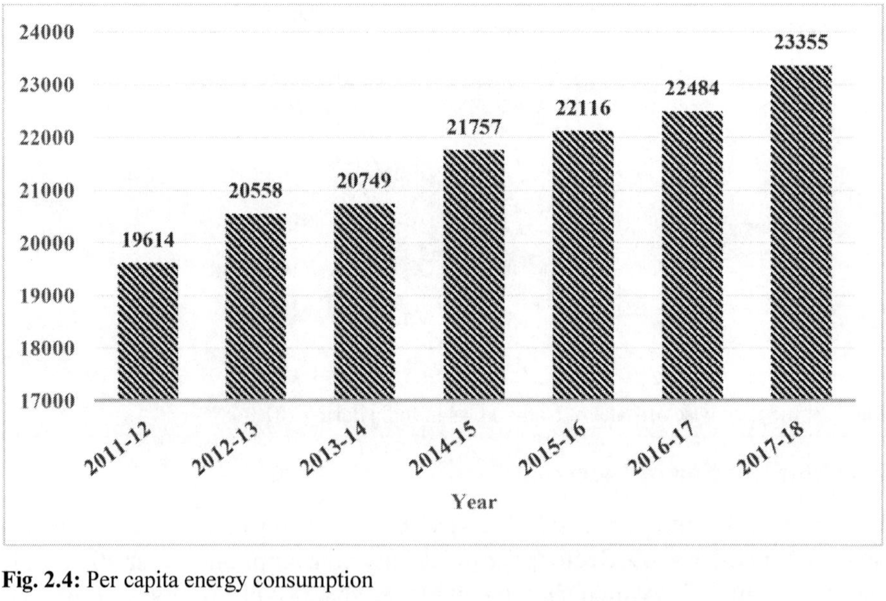

Fig. 2.4: Per capita energy consumption

3

Solar Energy

3.1 Introduction

The sun emits electromagnetic radiation, which is referred to as solar radiation. Every location on the Earth's surface receives some sunlight during the course of the year, but the amount of solar radiation varies. On a daily basis, we see much less energy from the sun than it emits and incident upon the earth. In an hour and a half, the amount of sunlight that reaches the surface of the earth is enough to provide enough energy for an entire year. Almost all electromagnetic radiation from the sun is invisible to us. Solar radiation often called solar resource or simply sunlight, is a general term for electromagnetic radiation emitted by the sun. Various technologies are available for capturing solar radiation and converting it into useful forms of energy, like heat and electricity. Technical feasibility and economic operation of these technologies depends on the solar resources available at a given location.

3.2 Solar Radiation

The sun is a hot sphere of gas whose internal temperatures reach over 20 million K due to nuclear fusion reactions at the sun's core which convert hydrogen to helium. The radiation from the inner core is not visible since it is strongly absorbed by a layer of hydrogen atoms closer to the sun's surface. Heat is transferred through this layer by convection. The surface of the sun, called the photosphere, is at a temperature of about 6000 K and closely approximates a blackbody. For simplicity, the 6000 K spectrum is commonly used in calculations. but temperatures of 5762 ± 50 K and 5730 ± 90 K have also been proposed as a more accurate fit to the sun's spectrum. Astronomers use 5778 K when classifying the sun as a star. Figure 3.1 shows the spectrum of solar radiation for ultraviolet, visible and infrared radiations and their respective wavelength.

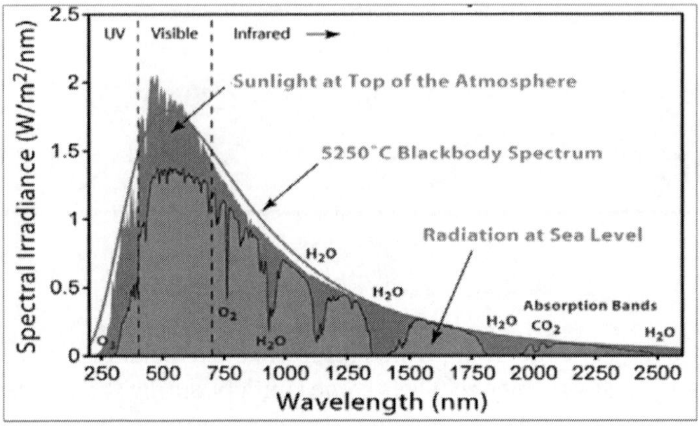

Fig. 3.1: Solar radiation spectrum

3.3 Solar Radiation at the Earth's Surface

Even though there is a relatively constant amount of solar radiation impacting Earth's atmosphere, the radiation hitting the surface of the planet has a wide range. Solar radiation received at the surface of the Earth is affected by several factors. These changes include variations in the overall power received, the spectral content of the light and the angle from which light is incident on a surface. Changing conditions can also affect the solar radiation at a particular location. In addition to local factors like clouds and seasonal variations, other factors like the length of the day at a particular latitude also contribute to the variability. Clouds and other local atmospheric phenomena tend to reduce variations in desert regions. The seasons are not highly variable in equatorial regions. The reason of variable solar radiations are-

a) Atmospheric effects, including absorption and scattering;
b) Local variations in the atmosphere, such as water vapour, clouds, and pollution;
c) Latitude of the location; and
d) The season of the year and the time of day.
e) Geographic location
f) Time of day
g) Season
h) Local landscape
i) Local weather

3.4 Atmospheric Effects

Atmospheric effects have several impacts on the solar radiation at the Earth's surface. The major effects for photovoltaic applications are:

a) By absorbing, scattering and reflecting the solar radiation in the atmosphere, solar radiation becomes less powerful

b) Absorption or scattering of some wavelengths of solar radiation results in a change in the spectral content of the radiation

c) An indirect or diffuse component to solar radiation

d) Additional effects on incident power, spectrum, and directionality are caused by local changes in the atmosphere (including water vapor, clouds, and pollution)

Earth's rounded shape enables the sun to strike the surface from different angles, starting at 0° (just above the horizon) and ending at 90° (directly overhead). Solar rays falling vertically on Earth's surface deliver the maximum amount of energy. When the sun's rays are slanted, they travel through the atmosphere for a longer period of time, becoming more scattered and diffuse in nature. As the Earth is round, the frigid polar regions seldom receive high sunshine, and due to the tilt of the earth's axis, they do not get any sunlight for part of the year. There are times of the year where the earth is closer to the sun than when it revolves around the sun in an elliptical orbit. Sunlight reaches the Earth's surface more readily when it is closer to the Sun. During the summer in the southern hemisphere and the winter in the northern hemisphere, the Earth is nearer the sun. Earth's tilt of 23.5° in its axis of rotation has a stronger impact on the amount of sunlight that strikes a particular location. From the vernal equinox in the spring to the autumnal equinox, tilting causes longer days in the northern hemisphere, and longer days in the southern hemisphere for the remaining six months (Figure 3.2). Each year, on or about March 23 and September 22, days and nights last exactly 12 hours.

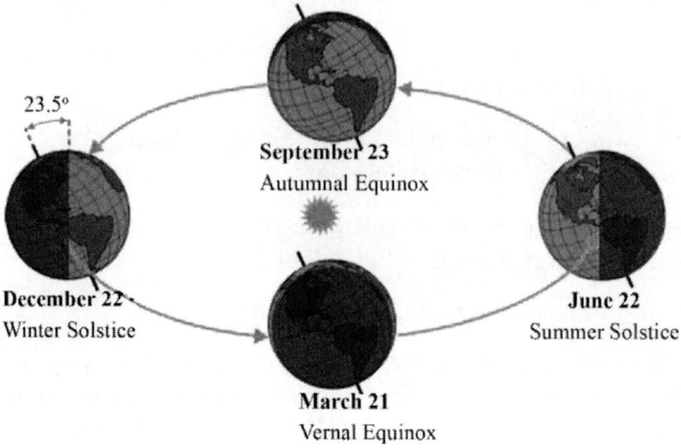

Fig. 3.2: Equinox and solstice

3.5 Solar Constant

Solar constant is the total radiation energy received from the Sun per unit of time per unit of area on a theoretical surface perpendicular to the Sun's rays and at Earth's mean distance from the Sun. The most accurate measurements are obtained from satellites, which are free of atmospheric effects. The value of the constant is approximately 1.370 kilowatts per square metre. The solar constant actually varies by +/-3% because of the Earth's slightly elliptical orbit around the Sun. Sunlight has a luminosity of 3.86 x 10^{26} watts. Most of this radiation is in the visible and infrared part of the electromagnetic spectrum, with less than 1% emitted in the radio, UV and X-ray spectral bands. Because the Sun is about 150 million kilometres from the Earth, and because the Earth is about 6300 km in radius, only 0.000000045% of this power is intercepted by our planet. This still amounts to a massive 1.75 x 10^{17} watts.

3.6 Air Mass

The Air Mass is the path length which light takes through the atmosphere normalized to the shortest possible path length (that is, when the sun is directly overhead) and shown in Figure 3.3. The Air Mass quantifies the reduction in the power of light as it passes through the atmosphere and is absorbed by air and dust. The Air Mass is defined as:

$AM=1/\cos\theta$

where θ is the angle from the vertical (zenith angle). When the sun is directly overhead, the Air Mass is 1.

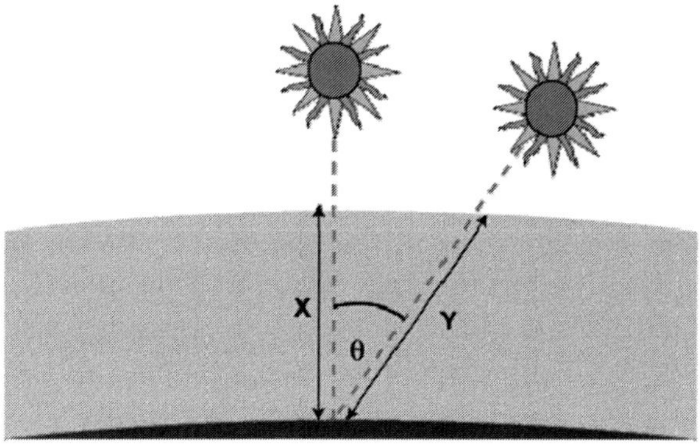

Fig. 3.3: Air mass

3.7 Absorption in the Atmosphere

Solar radiation is absorbed by gasses, dust, and aerosols as it passes through the atmosphere. Certain atmospheric gases, notably ozone (O_3), carbon dioxide (CO_2), and water vapor (H_2O), have high absorption of photons with energies close to their bonds. The spectral radiation curve is characterized due to this absorption. As an example, water vapor and carbon dioxide absorb a great deal of the far infrared light above 2 μm. The ozone also absorbs most ultraviolet light below 0.3 μm. Spectral content of the terrestrial solar radiation changes with the absorption of specific gasses in the atmosphere, but they have relatively little impact on the overall intensity. In reality, absorption does not cause deep troughs in spectral irradiance, but rather reduces power depending on path length through the atmosphere. A relatively uniform reduction in visible light is caused by these atmospheric elements, so when the sun is overhead, incident light appears white due to the absorption caused by these elements. In contrast, higher energy (lower wavelength) light is more effectively absorbed and scattered along longer paths. Hence in the morning and evening the sun appears much redder and has a lower intensity than in the middle of the day.

3.8 Type of Radiations

Different types of solar radiation can be understood from the Figure 3.4. Direct radiation is also known as beam radiation or direct beam radiation. This term describes solar radiation traveling from the sun down to the earth's surface in a straight line. A direct beam of radiation has a definite direction, so if all the rays are headed in the same direction, then an object can block them all at once. Because of this, shadows only appear when direct sunlight is blocked. Direct beam radiation can be reduced by 10% on clear, dry days, and by 100% on thick, cloudy days due to atmospheric conditions. Diffuse radiation is composed of sunlight scattered by atmospheric molecules and particles, but still reaching the earth's surface. Global solar radiation is the sum of diffuse and direct solar radiation.

Albedo

Albedo refers to the fraction of solar radiation reflected by the ground, the land cover, and water bodies on earth's surface. In addition to reflectance from clouds and air, astronomers and meteorologists consider atmospheric reflectance. Sometimes it is referred as ground reflectance.

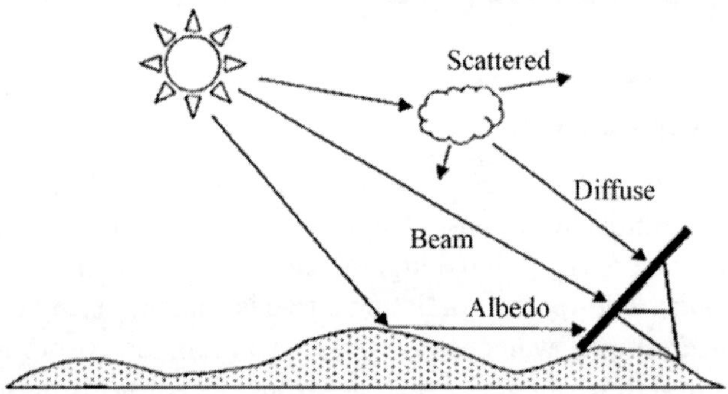

Fig. 3.4: Types of solar radiation

3.9 Solar Energy Technologies

Solar technologies capture this radiation and turn it into useful forms of energy. Solar energy can be harnessed in three ways: photovoltaics, solar heating, and solar concentrating. Photovoltaic panels generate electricity directly from sunlight through an electronic process, and can power small electronics such as calculators and road signs, as well as homes and large commercial enterprises. Heat generated by the sun is used in both solar heating and cooling (SHC) and concentrating solar power (CSP) applications to heat space or water, respectively, or to run traditional turbine generators, respectively. Thermal energy is useful for many process heat applications and for power generation via the Rankine cycle. Photovoltaic (direct conversion of solar energy into DC current) solar cell modules can also be used to generate electricity for household uses. Using photovoltaic (PV) panels or mirrors, solar energy systems convert sunlight into electricity. Photovoltaics is the direct conversion of light (photo) to electricity (-voltaic). In addition to converting this power into electricity, it can be stored thermally or in batteries. Using solar energy by direct conversion and indirect conversion is shown in Figure 3.5.

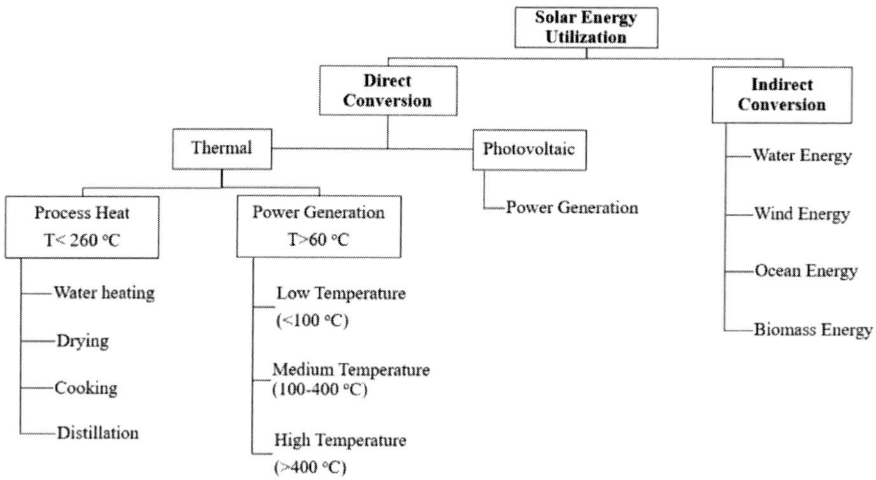

Fig. 3.5: Classification of solar energy utilization

3.10 Solar Thermal Application

A solar collector is a device that collects and/or concentrates solar radiation from the sun. Active solar heating is the primary purpose of these devices, and they allow for the heating of water for personal use. Solar collectors are generally mounted on roofs and must be extremely durable since they are exposed to a variety of weather conditions. Solar collectors provide an alternative to traditional domestic hot water heaters, potentially reducing energy costs in the long run. In addition to being used in domestic settings, solar thermal power plants can combine these collectors in arrays to generate electricity.

3.11 Types of Solar Collectors

Solar collectors come in many different types, but they all follow the same basic principles. Heat is generated by collecting and focusing energy from the Sun and using it to heat water. In the simplest of these devices, water flows through pipes surrounded by black material. Black materials absorb solar radiation very well, and as they heat up, they make the water around them hot. Despite this simple design, collectors can be very complicated. If a high-temperature increase isn't required, absorber plates can be used, but generally devices that use reflective materials to focus sunlight result in a higher temperature increase.

Flat Plate Collectors

Flat-plate solar collectors (Figure 3.6) are probably the most fundamental and extensively studied technology for solar-powered domestic hot water systems. Flat plate collectors were developed in the 1950s as a method of using solar energy to produce hot water for domestic use. During the heating process, water passes through a black plate under a transparent cover before returning to an insulated hot water storage cylinder. Heating domestic hot water is the primary use of flat plate collectors. Basically, this technology works on a very simple principle. As the Sun heats a dark flat surface, the energy is collected and transferred to water, air, or other fluids. An average flat-plate solar collector consists of the following components:

- Black surface - absorbent of the incident solar energy
- Glazing cover - a transparent layer that transmits radiation to the absorber, but prevents radiative and convective heat loss from the surface
- Tubes containing heating fluid to transfer the heat from the collector
- Support structure to protect the components and hold them in place
- Insulation covering sides and bottom of the collector to reduce heat losses

The collectors consist of metal boxes with transparent glazing over dark-coloured absorber plates. In order to minimize heat loss to other parts of the collector, the sides and bottom of the collector are usually insulated. Solar radiation passes through the transparent glazing material and hits the absorbing plate. This plate heats up, transferring the heat to either water or air held between the glazing and absorbing plate. The absorber plates are sometimes coated with special coatings that absorb and retain heat more effectively than traditional black paints. Generally, these plates are made from metal that is a good conductor, such as copper or aluminum.

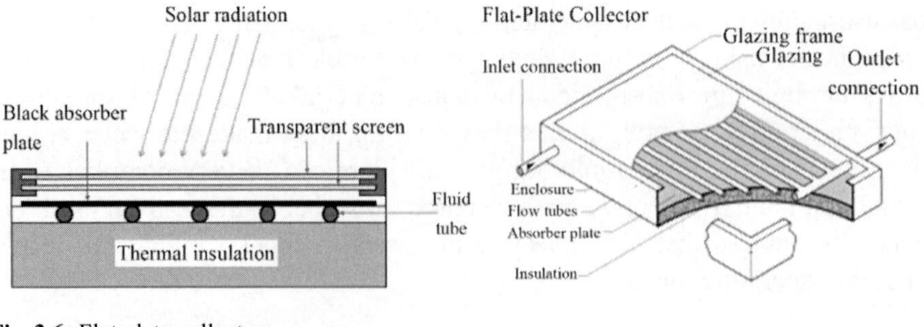

Fig. 3.6: Flat plate collector

It is commonly found for flat-plate systems to operate and achieve maximum efficiency in the temperature range from 30 to 80 °C, but some new types of collectors employing vacuum insulation can reach temperatures up to 100 °C. Selected coatings have enabled flat-plate collectors to reach 200°C when stagnant fluid is introduced. It is common to install flat-plate collectors facing the equator (i.e. south-facing in the Northern hemisphere, north-facing in the Southern hemisphere). Ideally, the collector plate should be tilted to match the latitude of the location (+/-15°). A solar cooling system should be installed at Latitude - 10°, so the solar beam is perpendicular to the collector during summer. It is recommended that solar heating be installed at an angle of Latitude + 10° if the application involves solar heating. Nevertheless, for year-round hot water applications, it was found that Latitude + 5° provides slightly better performance during winter, when hot water is needed most.

3.12 Transport Fluid Options

The flat plate collectors can involve liquid or air heat transport. *Water* is one of the common options as liquid fluid due to its accessibility and good thermal properties:

- It has a relatively high volumetric heat capacity
- It is incompressible (or almost incompressible)
- It has a high mass density (which allows using small tubes and pipes for transport)

In the winter, water can freeze and damage collectors and pipelines. It is possible to mitigate this effect by draining down the collector when solar inputs are low (below critical levels of insolation). Often, drain down sensors are used to monitor the system and ensure that pockets of water have been drained completely, in order to prevent damage from pocket water freezing. The system also needs to be refilled the following morning. It is possible that air pockets can form in the collector, blocking water flow and decreasing efficiency. To solve such problems, antifreeze mixtures can be substituted for pure water. Ethylene glycol and propylene glycol are the most common antifreeze components. Due to their toxicity, chemicals mixed with water require closed-loop systems and proper disposal. Antifreeze should be replaced every five years on average.

In some flat-plate collector designs, air can be used as a transport fluid. Space heating or crop drying are better suited to this option. It is usually necessary to have a fan in the system to facilitate air flow and ensure efficient heat transfer. Depending on the design, thermal buoyancy can provide passive (no fan) air movement.

It is also possible to use phase-change liquids with flat-plate collectors. This group of fluids includes some refrigerants. Due to their low boiling point, they can change from liquid to gas as the temperature increases, eliminating the problems described above for water. They do not freeze, so they do not cause the problems described above for water. Fluids with these properties can be useful in settings with rapid temperature fluctuations.

3.13 Collector Construction

Designing flat plate collectors requires maximizing absorption, minimizing reflection and radiation losses, and providing effective heat transfer between the collector plate and fluid. Heat-transfer fluids must be contained in tubes or ducts that form a good thermal bond with the absorber plate. In order to address this issue, different construction designs have been proposed as shown in Figure 3.7. The following color codes describe the material: light blue - glass, dark blue - fluid channels, black - absorb material, gray - insulation. To increase thermal conductivity between the components, some constructions incorporate fluid channels. Other modifications include tubes and channels soldered or cemented to the plate.

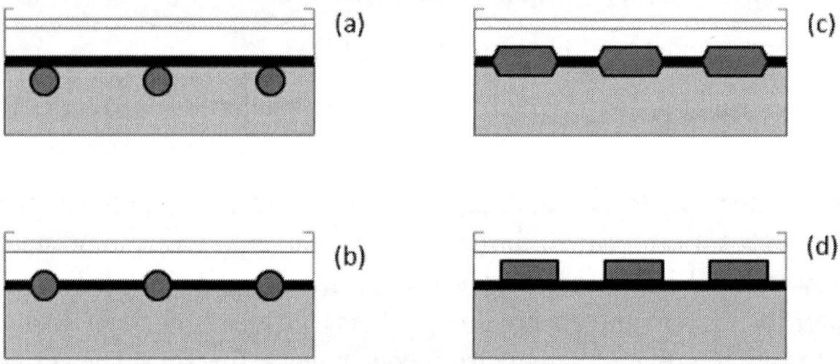

Fig. 3.7: Cross section of fluid transport tubes

Some advantages of the flat-plate collectors are that they are:

- Easy to manufacture
- Low cost
- Collect both beam and diffuse radiation
- Permanently fixed (no sophisticated positioning or tracking equipment is required)
- Little maintenance

3.14 Energy Balance in Flat-Plate Collectors

Energy balance calculations under steady state conditions are a fundamental concept in thermal analysis of any thermal system. It is the difference between absorbed solar radiation and the collector's total thermal losses that determines the useful energy output of a collector in steady-state.

Useful energy = Absorbed solar energy - Thermal losses

The higher the useful energy output of a particular design, the higher the expected efficiency. This type of analysis requires considering thermal efficiency of the collector as it provides a basis for comparing different materials and modifications of collector systems.

3.15 Flat Plate Collector Performance

The maximum possible useful energy gain can be achieved when the collector is at the same temperature as the inlet fluid. In this case, the heat losses are minimized. However, in an actual operation setting, this is not always the case. To describe the effective (actual) useful energy gain via heat exchange, we should introduce the *heat removal factor* - F_R

This coefficient shows how much energy remains after heat losses to the surrounding due to collector and inlet temperature difference. Therefore, the energy balance equation for the actual system can be written as follows-

Qu =AcFR[S−UL(Ti−Ta)]Qu

=AcFR[S−UL(Ti−Ta)]

This equation reminds the energy balance equation discussed earlier, only with the F_R factor. This flow factor depends on the mass flow rate of the fluid and heat capacity, and you can learn more details about the flow factor and practical application of the above equation from the following reading.

The basic method of assessment of collector performance is to expose the system to solar radiation, run the fluid through it, and measure the inlet and outlet temperature along with the flow rate. Then the useful energy gain can be calculated from the experimental data as follows

Qu = mCp(To − Ti)Qu = mCp (To−Ti)

In addition, the incident radiation on the collector (G_T) and ambient temperature (T_a) can be recorded, so we can express the useful gain in terms of incident radiation:

$$Qu = AcFR\,[GT(\tau\alpha)\ average - UL(To - Ti)]\ Qu$$
$$= AcFR\,[GT(\tau\alpha)\ average - UL(To - Ti)]$$

and further the experimental efficiency of the system at each instant of operation can be obtained:

$$\eta i = QuAcGT$$

Thermal efficiency (η)

$$\eta = QuAcGT\eta = QuAcGT$$

Where Q_u is the useful energy output from a collector, G_T is the incident solar radiation flux (irradience), and A_c is the collector area. So the denominator here is the total energy input for the collector. In this formula the G_T is the parameter characterizing the external conditions, and it is usually known from practical measurements (with a pyranometer) or assumptions for a specific location.

The energy balance can also be expressed via the following key equation:

$$Qu = Ac[S - UL(T_{plate} - T_{ambient})]Qu$$
$$= Ac[S - UL(T_{plate} - T_{ambient})]$$

where S is the absorbed solar radiation, U_L is the total losses, T_{plate} is the temperature of the absorbing plate, and $T_{ambient}$ is the temperature of the air, and A_c is the area of the collector surface.

In a general case, when measurements of incident solar radiation (I_T) are available, the convenient approximation for the absorbed energy is given by:

$$S = (\tau\alpha)_{av} ITS$$
$$= (\tau\alpha)_{av} I_T$$

where $(\tau\alpha)_{av}$ is the product of transmittance of the collector cover and absorptance of the plate averaged over different types of radiation. In fact, $(\tau\alpha)_{av} \approx 0.96(\tau\alpha)_{beam}$ based on practical estimations.

3.16 Concentrating Solar Collectors

Solar concentrators come in many configurations, from cylindrical ones that focus on lines (tubes) to circular ones that focus on points (power towers) which can be seen in Figure 3.8. In addition to increasing radiation flux on the receiver, these configurations increase the radiation flux by reflecting light through many angled or curved mirrors. Concentration of light on the receiver is achieved by shaping the reflectors (mirrors) around the receiver (blue circle)

Types of concentrating collectors:

(a) Tubular absorbers with a diffuse back reflector
(b) Tubular absorbers with specular cusp reflectors
(c) Plane receiver with plain reflectors (v-trough)
(d) Multisectional planar concentrator
(e) Compound parabolic concentrator
(f) Parabolic trough
(g) Fresnel concentrator
(h) Array reflectors (heliostats) with a central receiver

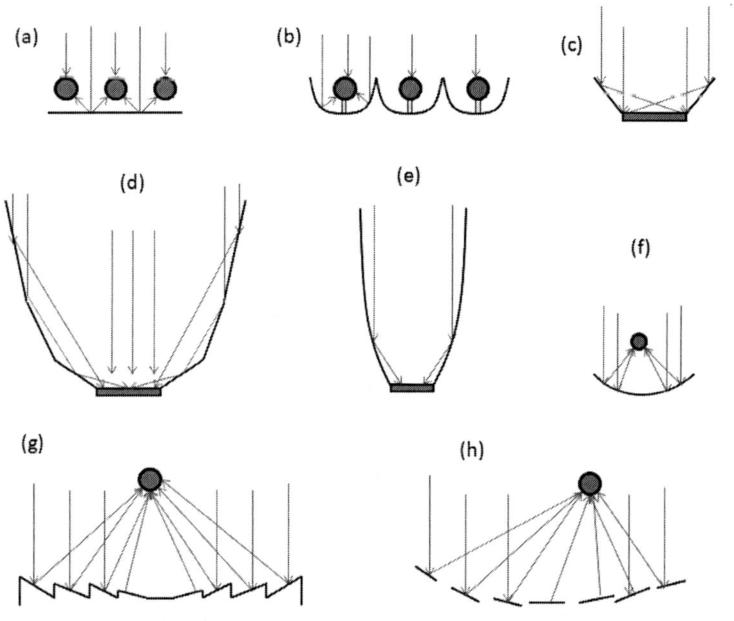

Fig. 3.8: Types of concentrating solar collectors

The light concentration process (Figure 3.9) is typically characterized by the concentration ratio (*C*). A concentration ratio is a factor by which incident energy flux (Io) is optically enhanced on receiving surface (Ir). It would be possible to increase the flux by concentrating the available energy coming through a chosen aperture on a smaller area of the receiver.

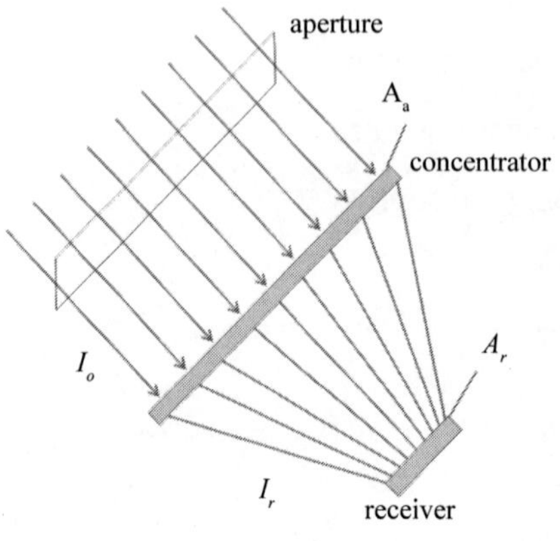

Fig. 3.9: Schematic representation of light concentration process

The ratio of area of the receiver to the area of the aperture, $C_{geo} = Ar/A_a$ is called the area concentration ratio (in some sources also called geometric concentration ratio). Although it is easy to use due to the known areas of the devices, it is only effective when the radiation flux is uniform over the aperture and receiver. Moreover, please note that the available receiver surface area can differ for some imaging concentrators from what the concentrator produces on the receiver surface. When the image does not cover the entire receiver surface, the concentration ratio must be derived from the area of the image. Energy flux ratios at the aperture and receiver can also be used to represent concentration ratios. This is referred to as flux concentration ratio C_{opt} (in some sources, optical concentration ratio) and can be applied directly to thermal calculations.

C_{opt} =Average flux over the receiver Flux over the aperture (insolation)

=1Ar∫IrDArIO

In case the ambient energy flux over the aperture (insolation) and over the receiver (irradiance) is uniform, the geometric and optical concentration ratios are equal ($C_{geo} = C_{opt}$).

A concentration ratio is an important metric for identifying and ranking optical concentrators. Solar concentration has a theoretical limit. Based on geometrical considerations, the limit for circular concentrators is 45,000, and for linear concentrators is 212; however, real systems may be unable to reach these limits due to non-idealities and losses. The concentration range of sunlight can be divided into three categories: low concentration range (C<10), medium concentration range (10<C<100), and high concentration range (C>100). A single concentration ratio is only possible with some of the systems (e.g., V-troughs and pyramidal plane reflectors). Conical, parabolic, and spherical systems have curved reflecting surfaces that produce a distribution of flux density over the receiver, rather than a variable C. To characterize the performance of such a concentrator, a local concentration ratio (Cl) is used:<100>

Cl=Local intensity Incident intensity=I(y)Iap

where I(y) is determined for any local position y from the center of the produced image, and I_{ap} is the intensity of the incident radiation at the aperture.

In many typical cases of imaging concentrators, the reflectance of the surface (ρ), i.e., the fraction of light radiation reflected from the surface compared to the total incident radiation, is also taken into account. Then the local intensity of the concentrated light, *I(y)*, can be described as follows:

I(y)=IapρCl

Solar Thermal Fluids

Solar thermal fluids (or heat-transfer fluides - HTF) come in six primary groups:

- Oil-based
- Water-based
- Molten salts
- Air
- Refrigerants
- Silicones

Solar thermal energy conversion systems use a variety of heat transfer fluids, each with advantages and disadvantages. In a Parabolic Trough and Linear Fresnel collector system, oil, water, or molten salts can all be used, while in a power tower system, only water and molten salts (oil is excluded) along with air can be used. The most common type of system installed worldwide is the parametric trough system, with 90% of systems installed by 2013. In flat plate and concentrating systems, refrigerants and silicones are rarely used. Oil-based fluids can be classified into three categories: synthetic hydrocarbons, paraffin hydrocarbons, and aromatic refined mineral oils. There are two types of water-based fluids: pure water and water and glycol mixtures, which contain ethylene or propylene glycol. Since molten salts are high-temperature fluids, they are only available in concentrating systems. An air supply consists mostly of nitrogen and oxygen and is used for a wide variety of applications, including drying processes in industry or heating of buildings at low. Today, solar thermal systems do not commonly use refrigerants, which are typically found in refrigerators, air conditioners, and heat pumps. There are synthetic and organic silicones available, but they aren't commonly used in solar thermal systems.

The thermal fluid should possess the following important properties

- **Maximum temperature** is the point at which the fluid begins to decompose or break down. This maximum temperature is most likely to occur in the hottest parts of a system and must therefore be carefully designed for these key points. Critical locations include the solar collectors' absorbing tubes, as well as the heaters and boilers.
- **Freezing temperature** is the coldest temperature before the fluid changes phase into a solid, and cannot be pumped, potentially causing irreversible damage to system components. In the case of each fluid, additional properties such as the expansion of the fluid when it freezes are important when determining the potential impact of freezing.
- **Density** is the mass per unit volume and is temperature dependent. Density can change drastically for some fluids as they heat up. For example, hydrocarbon oils can have a ~30% decrease in density when increased from 25 degrees C to 390 degrees C.
- **Steam pressure** is the pressure required to prevent the fluid from changing state to a gas. At higher pressures, the temperature at which the heat transfer fluid changes phase from a liquid to a gas (steam) is also higher.
- **Specific heat** is measured in units of energy (J) per mass (kg) temperature (K) and represents the amount of energy required to raise the temperature of the heat transfer fluid per unit mass. For example, if a fluid had a specific heat of 2300 kJ/kgK, that fluid would require the supply of 2300 kJ of energy to raise one kg of the fluid by one degree Kelvin (Celsius). This parameter is very useful for energy and mass balance calculations.
- **Enthalpy** is measured in units of energy (J) per mass (kg), representing the energy contained in the fluid under specific conditions such as temperature.
- **Viscosity** is a fluid's resistance to shear stress; essentially a measure of how much pumping force is required to move the fluid in a constrained environment. Viscosity can change greatly with temperature for some fluids. Lower viscosities are preferred to reduce pumping costs.

3.17 Solar Thermal Power Systems

In solar thermal power/electrical generation systems, solar radiation is collected and concentrated to produce high-temperature heat for generating electricity. Two components are involved in all solar thermal power systems: reflectors (mirrors) and receivers that capture and focus sunlight. Steam is usually produced by heating and circulating a heat-transfer fluid in the receiver. An electric generator produces electricity by converting steam into mechanical energy through a turbine. The tracking systems of solar thermal power systems keep the sun's rays focused on the receiver throughout the day as the sun moves throughout the sky. There are usually an array or field of collectors in solar thermal power plants that supply heat to the turbine and generator A general scheme for solar thermal power generation is shown in Figure 3.10. The solar collector uses a thermal fluid to transfer solar energy into heat, which is then converted into steam in a boiler. Heat engines convert steam into mechanical energy by using a turbine, while rejecting some heat. When electric power is desired, mechanical energy is fed into a generator, where it is converted into electricity. At each conversion step, we can expect some losses due to non-100% efficiency. For heat engines to work efficiently, flat-plate collectors deliver too low temperatures; therefore, concentrating collectors (e.g. parabolic systems) or evacuated tubular collectors are more suitable. The overall efficiency of the power conversion system is composed of the efficiency of the solar collectors (with parabolic troughs, max ~75%), the efficiency of the heat engine (~35%). Minus field losses, the typical average overall efficiency of solar trough thermal plants is around 15-20%.

The main configurations of solar thermal power systems include:

- Parabolic troughs
- Parabolic dish
- Central receiver systems (power tower)
- Solar updraft tower

There are three main types of concentrating solar thermal power systems:

- Linear concentrating systems, which include parabolic troughs and linear Fresnel reflectors
- Solar power towers
- Solar dish/engine systems

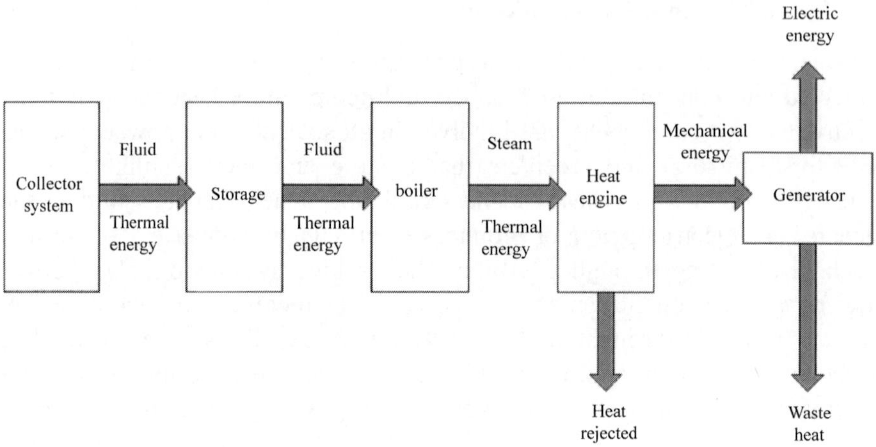

Fig. 3.10: Solar thermal power scheme

3.17.1 Parabolic Trough Collector Systems

A parabolic trough is (Figure 3.11) the most common solar thermal technology for utility-scale plants. These concentrating collectors have a very high potential for achieving fluid temperatures up to 500°C. A parabolic trough is a linear focus collector consisting of a cylindrically curved parabolic mirror, which reflects sunlight onto a tubular receiver in the focus line of the parabola. Steam is produced by absorbing heat in the tubular receiver and transferring it via circulation to the boiler or another device. Parallel rows of parabolic mirrors track the sun in a north-south direction or east-west direction based on location and energy production requirements, and are mounted on a north-south or east-west axis. Solar radiation is absorbed by the tubes and transferred to the heat exchange fluid passing through them. Fluid is pumped through the absorber tubes that are connected in series and parallel. It is common to use an insulated storage tank to generate power during times when the solar resource may be intermittent (as in cloud cover) or unavailable (e.g. in the early evening). If a storage tank exists, the heat transfer fluid is pumped to heat exchangers to transfer it to water (except in the case of direct steam generation, where water is already the heat transfer fluid and a heat exchanger is not required) in order to produce steam that can be expanded in a steam turbine to generate electricity.

The "trough" collectors that make up the solar field of a parabolic trough power plant are large cylindrical parabolic mirrors that concentrate the sunlight on a line of focus. Each of these collectors is installed in a row about a hundred meters long, resulting in a total solar field of about 500 meters and composed of many such parallel rows. Using mirrors, sunlight is focused on a metal absorber pipe more than 80 times. In order to reduce heat loss, this pipe is embedded in an evacuated glass tube. The surface of the absorber tube is coated with selective coating to reduce emission losses. Either water or a special thermal oil, runs through the absorber tube. Concentrated sunlight heats it up to nearly 400 degrees, converting it to steam. The steam is then condensed back into water, returning to the cycle after passing through the turbine.

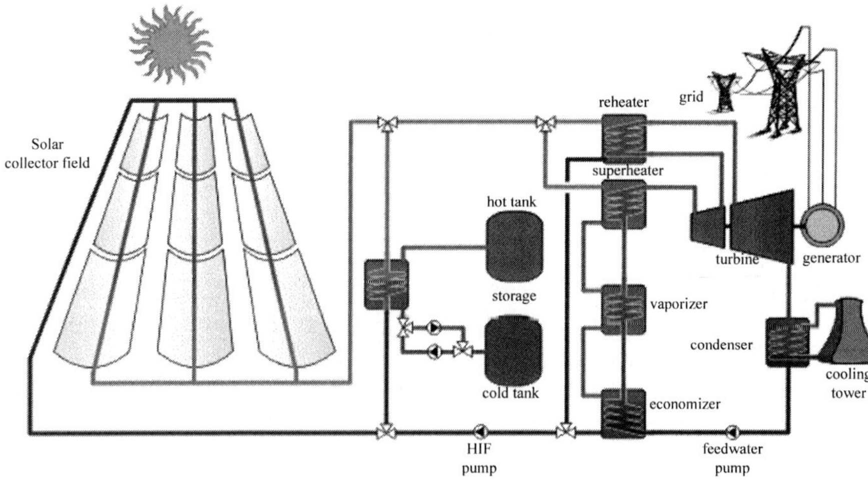

Fig. 3.11: Parabolic trough collector power generation

3.17.2 Central Receiver Systems - Power Tower

The central receivers are different from linear concentrators (troughs), which reflect concentrated light on a focal point as shown in Figure 3.12. A central receiver system is composed of five main components: heliostats, including their tracking system; receiver; heat transport and exchange; thermal storage; and controls. Solar power tower systems use heliostats to reflect and concentrate sunlight onto a receiver atop a tower with a large field of flat, sun-tracking mirrors. The intensity of sunlight can be increased by more than 1,500 times and achieve temperatures from 800 to over 1000°C. Water is used as a heat-transfer fluid in some power towers. Due to its superior heat transfer and energy storage capabilities, advanced designs are experimenting with molten nitrate salt. When the weather is cloudy or at night, the thermal energy storage system can produce electricity.

In such a system, multiple tracking mirrors (heliostats) are positioned in the field around the main external receiver, which is located on a tower. The concentration level of such systems is significantly higher than that of linear systems. A molten salt storage can be used to store the generated thermal energy, or it can be used to generate steam and then turn it into electricity (just like a regular power plant). Heat energy from salt is converted to steam by a conventional steam generator at the foot of the tower, which generates electricity. In addition to being able to store heat efficiently, molten salt can also be used to generate electricity for long periods of time. When power is needed from the plant, the hot salt is pumped to a generator that produces steam. The steam activates a turbine-generator system that creates electricity. From the steam generator, the salt is returned to the cold storage tank, where it is stored and can be eventually reheated in the receiver. A cylindrical receiver works best with Rankine cycle engines in large heliostat fields. As another option, Brayton cycle turbines, which operate at higher temperatures (about 1000°C), can be used; in this case, cavity receivers with a larger tower height to heliostat field area ratio would be preferable.

Central receiver systems are typically large-scale plants that are usually built to power a steam cycle. All energy can be collected at one location and transport networks can be saved due to the receiver's central location. In addition, heliostats always point at a different angle from the direct beam, so they harvest less energy compared to a parabolic concentrator because of their fixed position. As a result, the size of the collecting field must be increased to achieve the needed efficiencies of light concentration, which involves issues such as land use, higher environmental impact, and higher capital expenditures. Therefore, deserts and flat arid areas, which have plenty of sunlight and a low land value compared to other applications and industries, have a great deal of potential to develop large-scale central receiver solar plants. The power tower plants are quite large, generally 10 MWe or more, while the optimum sizes lie between 50-400 MW.

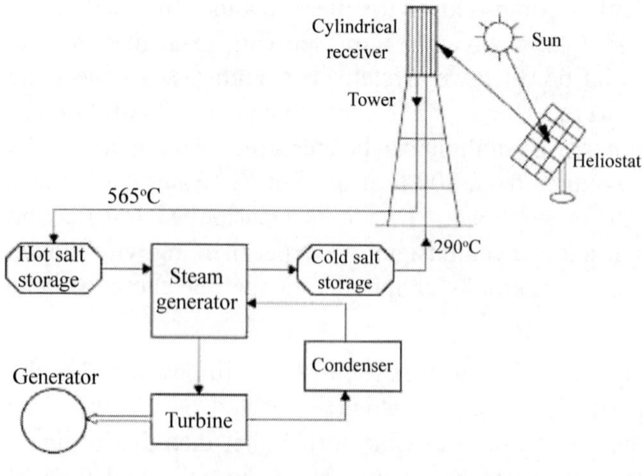

Fig. 3.12: Power Tower

3.17.3 Solar Chimney

Solar chimneys and solar updraft towers have been used to generate electric power for more than a century. It has taken several decades to develop several prototypes, and several of them have been implemented and operated. These prototypes vary in size and scale, with the largest ones capable of producing tens of kilowatt of power with towers that are a couple of hundred meters tall. There are three main parts to solar chimney power plants, despite their large scale. The collector transfers solar radiation to the system, the high chimney creates pressure differences, and the turbine produces power. Towers or chimneys are positioned at the center of the collector. In the system, the chimney acts like a motor, accelerating upward the system's air, which heats up in the collector and increases its kinetic energy. There is no need for any additional mechanism.

Working Principle

The working principle of solar chimney power plants is simple and fundamental even though they can be seen as large-scale complex structures. This system relies on the upward buoyancy of fluids based on their density difference and their movements, which are based on their pressure difference as shown in Figure 3.13. Semi-permeable collectors, which are exposed to the sun, transfer solar radiation falling on them due to their semi-permeable properties. When it is transmitted to the ground, the mechanism begins to work. Since the collector acts as a cover, solar radiation transfers to the system and heats the air. Air under the collector moves towards the collector due to the decrease in density of the heated air. Due to the opaque structure of the semi-permeable collector, the solar radiation reaching the ground causes a rise in temperature. Under the collector, the system's air is exposed to upward buoyancy because of the temperature difference on the ground. Below the collector, these forces push the air to the center, where it can only move. Long chimneys placed vertically in the center of the collector create a vacuum effect on the air as a result of the pressure difference at their entrance and exit. This causes the chimney to draw up the system's air. The system's air kinetic energy increases as a result of all of these effects, leading to an increase in its temperature and speed. At a certain height from the ground, there is a turbine in the chimney. When the air hits the turbine blades, the kinetic energy is converted to electrical energy, and power is generated.

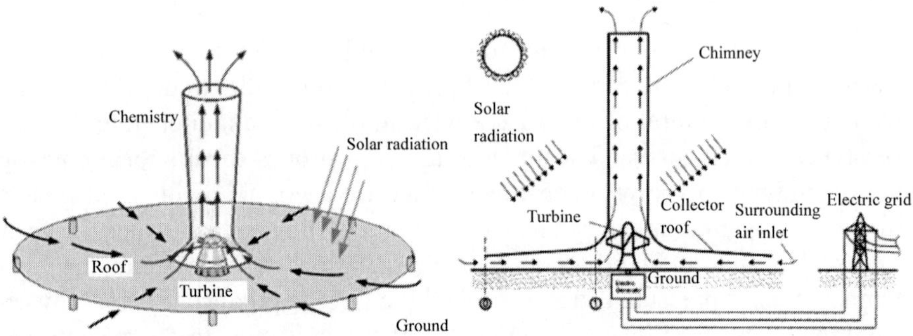

Fig. 3.13: Solar chimney

3.18 Solar Photovoltaic

A photovoltaic system (PV) converts light (photons) into electricity (voltage) through the photovoltaic effect. Solar photovoltaic (PV) power generation is the process of converting energy from the sun into electricity using solar panels. Solar cells made from silicon were first created by Bell Laboratories in 1954, making it possible to generate electricity by exposing them to sunlight. Satellites and smaller items like calculators and watches soon used solar cells to power their batteries. Solar cells, also known as photovoltaic cells, absorb, reflect, or pass through light when it incident on them. A PV cell is made of semiconductor material, and semiconductor material is better at conducting electricity than insulators but not as well as metals. PV cells use several types of semiconductor materials. Many countries have become cost competitive with solar energy and are deploying large-scale photovoltaic systems to power electricity grids.

As the semiconductor absorbs light energy, electrons in the material become negatively charged and transfer it to positively charged particles. Electrons are able to flow through this material as an electrical current because of this extra energy. A solar cell extracts this current through its conductive metal contacts, or grid-like lines, and uses it to power appliances and utility. Photovoltaic semiconductors have a bandgap, which indicates what wavelength of light they can absorb and convert to electricity. A semiconductor whose bandgap matches the wavelengths of incident light on a PV cell can efficiently utilize all the energy available to it. A PV system consists of arrays of solar panels, also called PV panels. PV systems can also be installed in grid-connected or off-grid (stand-alone) configurations. A PV cell's efficiency indicates how effectively it converts energy from one form to another, based on the amount of electrical power it produces in comparison to the amount of energy it receives from the light shining on it. A PV cell's ability to produce electricity depend on the light characteristics it receives (such as intensity and wavelength).

Working principle of PV

During the exposure of semiconductor materials to light, some of the photons of light rays are absorbed by the semiconductor crystals, which results in a significant number of free electrons. The most widely used semiconductor material for photovoltaic cells is silicon. Silicon atoms have four valence electrons. Each silicon atom in a solid crystal shares all four of its valence electrons with another atom nearby, which leads to covalent bonds between them. As a result, silicon crystals form tetrahedral lattices. A crystal absorbs sufficient photons from incident light to excite some of the electrons in its covalent bonds when the intensity of incident light is high enough. These excited electrons then get sufficient energy to migrate from valence band to conduction band. The electrons leaving the covalent bond as they are in the conduction band leave a hole behind each electron removed from the bond. As a result of the crystal structure of silicon, these free electrons move at random inside. Photovoltaic cells generate electricity by using these free electrons and holes.

Adding pentavalent impurity like phosphorus create a free electron in absence of formation of covalent bond between silicon atom. These free e;lectron conduct the current in semiconductor material.Impurities are introduced into semiconductors through the process of doping, and those impurities are called dopants. A pentavalent dopant that donates its fifth electron to a semiconductor crystal is called a donor. A semiconductor doped by donor impurities is called a negative semiconductor or an n-type semiconductor since the electrons are normally negatively charged.

A semiconductor crystal will produce the opposite type of semiconductor when trivalent impurities like boron are added instead of pentavalent phosphorous atoms. As a result, three complete covalent bonds will be created when three valance electrons of boron atom pair up with three valance electrons of silicon atoms. Incomplete bonds are formed when the fourth valence electron of silicon atoms is left unpaired. Hence, an incomplete bond always attracts electrons to fill the hole created by the absence of one electron. There is therefore a vacancy for the electron to occupy.The vacancy is known as a positive hole. There are considerable covalent bonds that are continually broken in a trivalent impurity doped semiconductor to complete other incomplete covalent bonds. The trivalent impurities called acceptor dopants can accept electrons for each hole, making them semiconductors doped with them p-type or positive type semiconductors.

Solar cells (also called photovoltaic cells or PV cells) convert light energy into electrical energy through the photovoltaic effect. Solar cells are essentially diodes with p-n junctions (Figure 3.14). Solar panels are made up of individual solar cells that can be combined to form modules. Solar cells with just one junction can produce a maximum open-circuit voltage of roughly 0.5 to 0.6 volts. Solar panels can generate considerable amounts of renewable energy when combined into large panels.

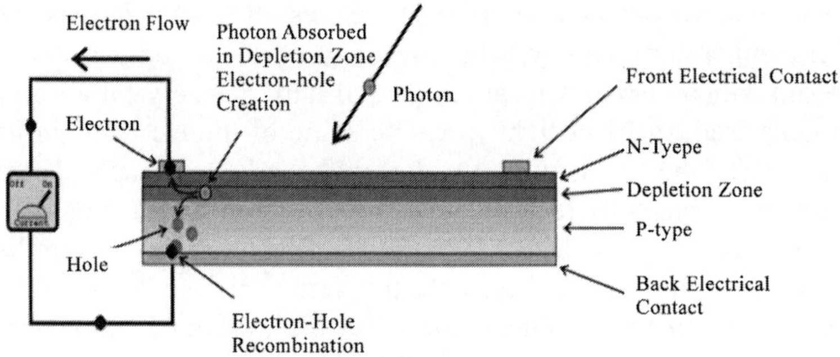

Fig. 3.14: Working of solar PV

3.19 Cells, Modules, Panels and Arrays

The photovoltaic cells are electrically connected in parallel or series circuits so that higher voltages, currents, and power levels can be produced. Photovoltaic modules consist of PV cell circuits sealed in an environmentally protective laminate, and are the fundamental building blocks of PV systems. Photovoltaic panels include one or more PV modules assembled as a pre-wired, field-installable unit. A photovoltaic array is the complete power-generating unit, consisting of any number of PV modules and panels Figure 3.15 shows the difference between photovoltaic cell, module, panel and array.

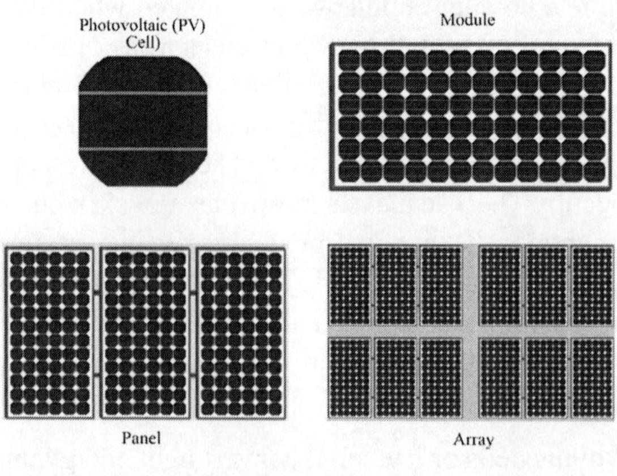

Fig. 3.15: Solar cell, module, panel and array

3.20 Advantages and Disadvantagees of PV System

Advantages

- Sunlight is free and readily available in many areas of the country.
- PV systems do not produce toxic gas emissions, greenhouse gases, or noise.
- PV systems do not have moving parts.
- PV systems reduce dependence on oil.
- PV systems have the ability to generate electricity in remote locations that are not linked to a grid.
- Grid-connected PV systems can reduce electric bills

Disadvantages

- PV systems have a high initial investment
- PV systems require large surface areas for electricity generation.
- The amount of sunlight can vary.
- PV systems require excess storage of energy or access to other sources, like the utility grid, when systems cannot provide full capacity.

3.21 Types of Solar Cell

Silicon

About 95% of today's solar cells are made from silicon, which is the most common semiconductor material. Cells made of crystallized silicon are composed of silicon atoms that are connected to one another. In this lattice, light is converted into electricity more efficiently due to its organized structure. Currently, silicon solar cells provide high efficiency, low cost, and long lifespan. It is expected that modules will last for 25 years or more, still producing more than 80% of their original power after this period.

Thin-film photovoltaics

In thin-film solar cells, one or more thin layers of PV material are deposited on a supporting material such as glass, plastic, or metal. Copper indium gallium diselenide (CIGS) and cadmium telluride (CdTe) are two of the most commonly used thin-film PV semiconductors today. The second-most common PV material after silicon is CdTe, and the material can be manufactured using low-cost methods. While this makes them a cost-effective alternative, their efficiencies still aren't quite as high as silicon.

Copper-Indium Gallium Diselenide (CIGS)

Copper-Indium Gallium Diselenide (CIGS) are made either by vapour deposition, or by "selenising" copper-indium films. Cadmium Telluride (CdTe). Cadmium telluride (CdTe) is made from the II-VI group elements, and has a direct band gap of 1.44 eV, making it one of the best-suited materials for photovoltaic applications.

Dye-Sensitized Solar Cells

A solar cell that uses liquid dyes is actually quite similar to a battery. At either end, electrodes are used, and one substance loses electrons while another gains them (oxidation and reduction, also known as redox). A solar cell differs only in that electrons are lost after a photon is absorbed (into the conduction band).

Perovskite photovoltaics

Perovskite solar cells have a unique crystal structure and are a type of thin-film cell. Printed, coated, or vacuum-deposited layers of materials are used to build perovskite cells. Layers of materials are applied to the substrate as a foundation. Their efficiency levels are similar to those of crystalline silicon, and they are generally easy to assemble. The efficiency of perovskite solar cells in the lab has increased by over 25% since 2009, far exceeding any other PV material.

Organic photovoltaics

Cells made with organic compounds, known as organic PV (OPV), can be tailored to enhance a specific function of the PV cell, such as transparency, color, or bandgap. While OPV cells are currently only about half as efficient as crystalline silicon cells and have a shorter lifetime, they could be less expensive to manufacture in large quantities.

Graphene Solar Cells

An organic material with a particularly high research interest is graphene, which is used in solar cells. A graphene sheet is a two-dimensional honeycomb of carbon with alternating double bonds.

3.22 I-V Characteristic

An I-V characteristic curve (Figure 3.16) shows the current and voltage characteristics of a particular photovoltaic (PV) cell, module, or array. It describes how it converts solar energy and how efficient it is. In order to determine a solar cell's output and solar efficiency, it is critical to know its electrical I-V characteristics (more specifically, its Pmax). An example of a silicon PV cell's current-voltage (I-V) characteristics is shown in the graph. Single solar cells or panels deliver power by combining their output current and voltage (I x V). Multiplying the voltages from short-circuit to open-circuit conditions, point by point, gives the power curve above.

Whenever a solar cell is open-circuited, i.e. not connected to any load, the current will be zero and the voltage across the cell will be its maximum. This is called the open-circuit voltage, or Voc, of the cell. On the other hand, when a solar cell is short circuited, which is by connecting its positive and negative leads together, it has its minimum voltage (zero) but its maximum current flowing out, referred to as the solar cell short circuit current or Isc. When a cell is in open circuit, it will provide the highest voltage, while when it is in closed circuit, it will deliver the highest current. The power reaches its maximum value at Imp and Vmp for a particular combination of current and voltage. A green rectangle indicates the top right area where maximum electrical power is generated by the cell. It is called the "maximum power point" or MPP. Therefore the ideal operation of a photovoltaic cell (or panel) is defined to be at the maximum power point. Solar cells have their maximum power point (MPP) near the bend in their I-V characteristic curves. From the open circuit voltage and short circuit current, we can estimate the values of Vmp and Imp: Vmp ≅ (0.8–0.90)Voc and Imp ≅ (0.85–0.95)Isc.

With changes in ambient temperature, solar cell output voltage and current will change, resulting in varying output power. An array of photovoltaic panels is made up of smaller PV panels that are interconnected. therefore, PV array I-V curves are essentially scaled up versions of single solar cell I-V characteristic curves.

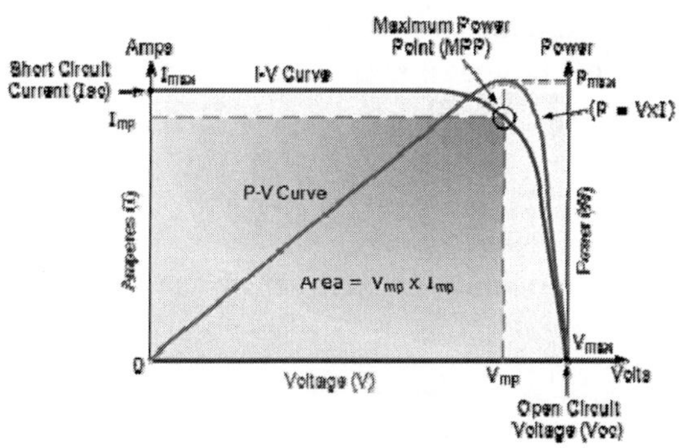

Fig. 3.16: I-V Characteristic of solar PV

3.23 Series and Parallel Connection of Solar Cell

The voltage or current capacity of the solar array can be increased by wiring or connecting together solar panels in series or parallel as shown in Figure 3.17. Voltage increases if the array panels are connected in series, while current increases if they are connected in parallel. The electrical power in Watts, generated by these different photovoltaic combinations will still be the product of the voltage times the current, (P = V x I). Solar arrays always have a maximum power point (MPP) in their upper right hand corner, regardless of how they are connected.

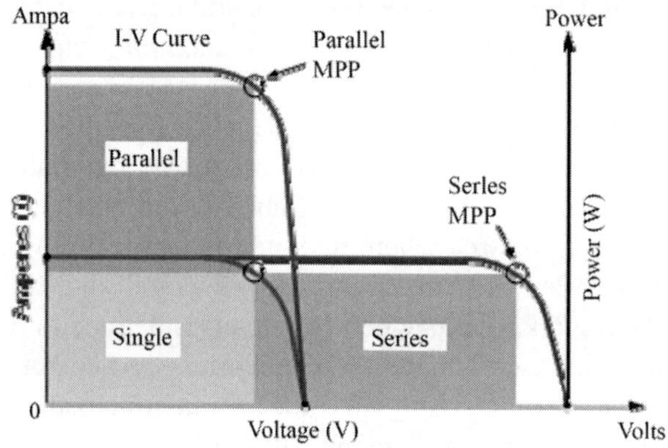

Fig. 3.17: Output of series and parallel connection

3.24 Electrical Characteristics of a Photovoltaic Array

Current and voltage are the electrical characteristics of a photovoltaic array. A PV array's output voltage (V) is affected by the solar insolation (solar irradiance) and the operating temperature of the solar cells. Generally, solar panel manufacturers provide I-V curves that summarize the relationship between current and voltage. The maximum current density or short circuit current density rating is just a ratio of the maximum current to the exposed surface area of the cell.

$$Fill\ Factor = \frac{P_m}{I_{sc} \times V_{oc}}$$

Where, I_{sc} is short circuit current, J_{sc} maximum current density and A is the area of solar cell.

Solar Array Parameters

- V_{OC} = open-circuit voltage

Array output voltage when no load is attached to the terminals (open circuit). PV array operations are fixed by the load, so this value is much higher than Vmp. Depending on how many PV panels are connected in series, this value will vary.

• I_{SC} = short-circuit current

A PV array's maximum current when its output connectors are shorted together (a short circuit condition). The value here is much higher than Imp, which corresponds to the circuit current normal for operation.

• MPP = maximum power point

This relates to the point where the power supplied by the array that is connected to the load (batteries, inverters) is at its maximum value, where MPP = Imp x Vmp. The maximum power point of a photovoltaic array is measured in Watts (W) or peak Watts (Wp).

• FF = fill factor

The fill factor is the relationship between the maximum power that the array can actually provide under normal operating conditions and the product of the open-circuit voltage multiplied by the short-circuit current, (V_{OC} x I_{SC}) This fill factor value gives an idea of the quality of the array and the closer the fill factor is to 1 (unity), the more power the array can provide. Typical values are between 0.7 and 0.8.

The ratio between product of current and voltage at maximum power point to the product of short circuit current and open circuit voltage of the solar cell.

$$Fill\ Factor = \frac{P_m}{I_{sc} \times V_{oc}}$$

• %eff = percent efficiency

The efficiency of a photovoltaic array is the ratio between the maximum electrical power that the array can produce compared to the amount of solar irradiance hitting the array. The efficiency of a typical solar array is normally low at around 10-12%, depending on the photovoltaic type (monocrystalline, polycrystalline, amorphous or thin film) of cell being used.

3.25 Efficiency of Solar Cell

It is defined as the ratio of maximum electrical power output to the radiation power input to the cell and it is expressed in percentage. It is considered that the radiation power on the earth is about 1000 watt/square metre hence if the exposed surface area of the cell is A then total radiation power on the cell will be 1000 A.watts. Hence the efficiency of a solar cell may be expressed as

$$Efficiency(\eta) = \frac{P_m}{P_{in}} \approx \frac{P_m}{1000A}$$

3.26 Solar PV plant

On-grid Solar Power System (Grid interactive System)

A grid-connected PV system (Figure 3.18) is most commonly used because it is easier to design and is typically less expensive than an off-grid PV system. Using grid-connected PV systems, homeowners can consume less power from the grid and supply unused or surplus energy back to the utility grid. A system's configuration and size are determined by its application. Typically, grid-connected PV systems for residential use are rated under 20kW, those for commercial use are rated between 20kW and 1MW, and those for utility energy storage are rated above 1MW. To be used in a domestic building, direct current (DC) must be converted into alternating current (AC); this is accomplished by an inverter. Electricity is then passed from the generation meter, which measures how much electricity has been created, to the consumer unit, where it can be used within the property or exported back to the national grid.

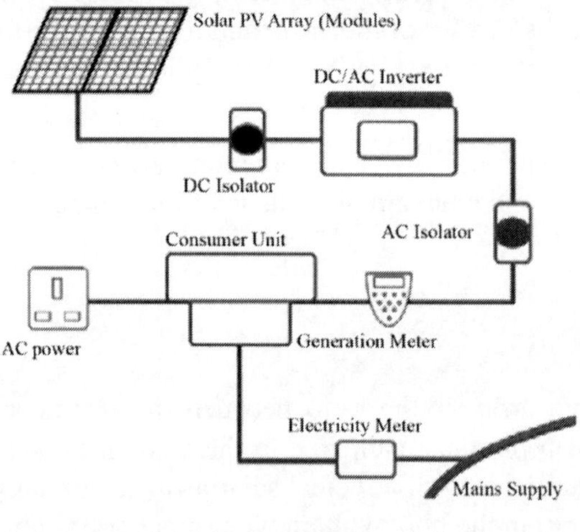

Fig. 3.18: On-grid Solar Power System

Off-grid Solar Power System (Standalone System)

Off-grid solar power system (Figure 3.19) is used when electricity/grid is not available or when a backup power supply is required. The solar panel generates energy, which is stored in the batteries connected to this system. When the grid is unavailability or when there are no sunlight hours, some of the most essential equipment and appliances are directly connected to this system.

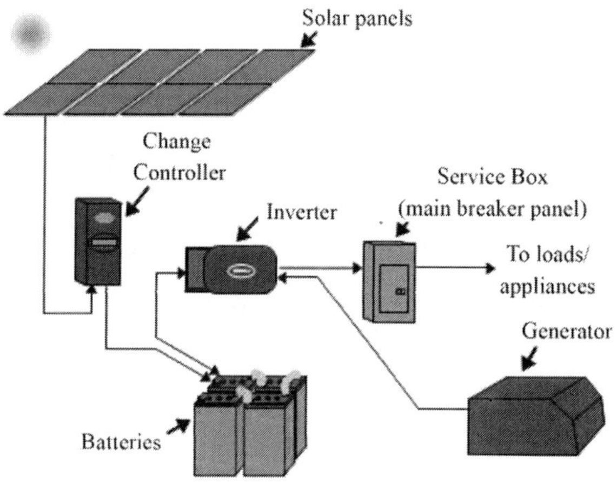

Fig. 3.19: Off-grid Solar Power System

Hybrid Solar Power System

Hybrid Solar Power System (Figure 3.20) combines off-grid and on-grid solar power. The excess energy generated by the solar system will be stored in the batteries connected to the system if the electricity consumption of your house is lower than the energy generated by the solar system. Batteries will be used if the house electricity consumption is greater than the solar energy generation. The excess energy generated through solar panels and batteries is used to power the grid/electricity department if the house electricity consumption is less than the energy generated through solar panels and batteries.

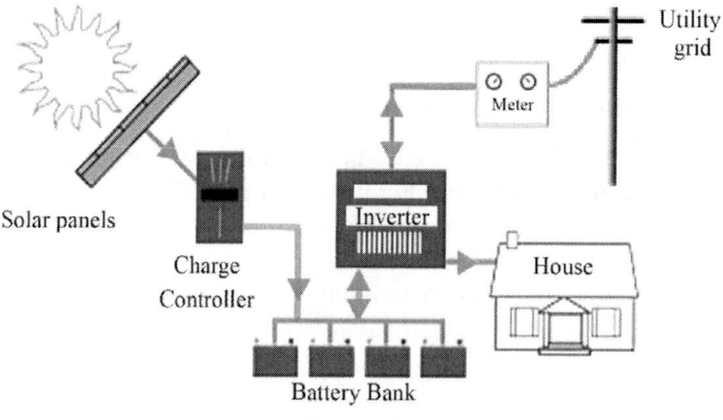

Fig. 3.20: Hybrid Solar Power System

3.27 Solar PV System Sizing

1. Determine power consumption demands

The first step in designing a solar PV system is to find out the total power and energy consumption of all loads that need to be supplied by the solar PV system as follows:

1a. Calculate total Watt-hours per day for each appliance used.

Add the Watt-hours needed for all appliances together to get the total Watt-hours per day which must be delivered to the appliances.

1b. Calculate total Watt-hours per day needed from the PV modules

Multiply the total appliances Watt-hours per day times 1.3 (the energy lost in the system) to get the total Watt-hours per day which must be provided by the panels.

2. Size the PV modules

Different size of PV modules will produce different amount of power. To find out the sizing of PV module, the total peak watt produced needs. The peak watt (Wp) produced depends on size of the PV module and climate of site location. We have to consider panel generation factor which is different in each site location. For India, the panel generation factor is 4-5. To determine the sizing of PV modules, calculate as follows:

2a. Calculate the total Watt-peak rating needed for PV modules

Divide the total Watt-hours per day needed from the PV modules (from item 1.2) by 4-5 to get the total Watt-peak rating needed for the PV panels needed to operate the appliances.

2b. Calculate the number of PV panels for the system

Divide the answer obtained in item 2.1 by the rated output Watt-peak of the PV modules available to you. Increase any fractional part of result to the next highest full number and that will be the number of PV modules required. Result of the calculation is the minimum number of PV panels. If more PV modules are installed, the system will perform better and battery life will be improved. If fewer PV modules are used, the system may not work at all during cloudy periods and battery life will be shortened.

3. Inverter sizing

An inverter is used in the system where AC power output is needed. The input rating of the inverter should never be lower than the total watt of appliances.

The inverter must have the same nominal voltage as your battery. For stand-alone systems, the inverter must be large enough to handle the total amount of Watts you will be using at one time. The inverter size should be 25-30% bigger than total Watts of appliances. In case of appliance type is motor or compressor then inverter size should be minimum 3 times the capacity of those appliances and must be added to the inverter capacity to handle surge current during starting. For grid tie systems or grid connected systems, the input rating of the inverter should be same as PV array rating to allow for safe and efficient operation.

4. Battery sizing

The battery type recommended for using in solar PV system is deep cycle battery. Deep cycle battery is specifically designed for to be discharged to low energy level and rapid recharged or cycle charged and discharged day after day for years. The battery should be large enough to store sufficient energy to operate the appliances at night and cloudy days. To find out the size of battery, calculate as follows-

4a. Calculate total Watt-hours per day used by appliances

4b. Divide the total Watt-hours per day used by 0.85 for battery loss

4c. Divide the answer obtained in item 4.2 by 0.6 for depth of discharge

4d. Divide the answer obtained in item 4.3 by the nominal battery voltage

4e. Multiply the answer obtained in item 4.4 with days of autonomy (the number of days that need the system to operate when there is no power produced by PV panels) to get the required Ampere-hour capacity of deep-cycle battery-

$$\text{Battery Capacity (Ah)} = \frac{\text{Total Watt-hours per day used by appliances} \times \text{Days of autonomy}}{(0.85 \times 0.6 \times \text{nominal battery voltage})}$$

5. Solar charge controller sizing

The solar charge controller is typically rated against Amperage and Voltage capacities. Select the solar charge controller to match the voltage of PV array and batteries and then identify which type of solar charge controller is right for your application. Make sure that solar charge controller has enough capacity to handle the current from PV array.

For the series charge controller type, the sizing of controller depends on the total PV input current which is delivered to the controller and also depends on PV panel configuration (series or parallel configuration).According to standard practice, the sizing of solar charge controller is to take the short circuit current (Isc) of the PV array, and multiply it by 1.3

Solar charge controller rating = Total short circuit current of PV array x 1.3

Example: A house has the following electrical appliance usage:

- One 18 Watt fluorescent lamp with electronic ballast used 4 hours per day.
- One 60 Watt fan used for 2 hours per day.
- One 75 Watt refrigerator that runs 24 hours per day with compressor run 12 hours and off 12 hours.

The system will be powered by 12 Vdc, 110 Wp PV module.

1. **Determine power consumption demands**

Total appliance use

= (18 W x 4 hours) + (60 W x 2 hours) + (75 W x 24 x 0.5 hours)

= 1,092 Wh/day

Total PV panels energy needed

= 1,092 x 1.3

= 1,419.6 Wh/day

2. Size the PV panel

Total Wp of PV panel capacity needed

= 1,419.6 / 3.4

= 413.9 Wp

Number of PV panels needed

= 413.9 / 110

= 3.76 modules

Actual requirement = 4 modules

So this system should be powered by at least 4 modules of 110 Wp PV module

3. Inverter sizing

Total Watt of all appliances = 18 + 60 + 75 = 153 W

For safety, the inverter should be considered 25-30% bigger size.

The inverter size should be about 190 W or greater

4. Battery sizing

Total appliances use

= (18 W x 4 hours) + (60 W x 2 hours) + (75 W x 12 hours

Nominal battery voltage = 12 V

Days of autonomy = 3 days

Battery capacity = [(18 W x 4 hours) + (60 W x 2 hours) + (75 W x 12 hours)] x 3
(0.85 x 0.6 x 12)

Total Ampere-hours required 535.29 Ah

So the battery should be rated 12 V 600 Ah for 3 day autonomy

5. Solar charge controller sizing

PV module specification

Pm = 110 Wp

Vm = 16.7 Vdc

Im = 6.6 A

Voc = 20.7 A

Isc = 7.5 A

Solar charge controller rating = (4 strings x 7.5 A) x 1.3 = 39 A

So the solar charge controller should be rated 40 A at 12 V or greater.

4

Wind Energy

4.1 Introduction

The wind is caused by the motion of air. In different regions, there are pressure differences that cause winds. There is a tremendous amount of energy carried by them. The use of wind energy can be profitable in regions with strong winds for a significant part of the year. The wind is a form of solar energy. The wind is caused by uneven heating of air and land by the sun. The earth's rotation and geological irregularities also play a role. Heating causes air to rise, and cooling causes cooler air to fall to fill the space. As a result, the air mass moves. On a global basis, the poles exert a primary forcing function that allows the equator to receive surface winds from the poles. The low-density heated air is buoyed up by the equator. Air flowing from the poles displacements it at the surface by being cooler and denser. As a result, air near the equator tends to flow back toward the poles, not toward it. The north hemisphere is thus characterized by a north-to-south convective circulation with surface winds. The wind has been used as a source of energy for centuries. Small and large-scale wind turbines convert wind energy into electricity for personal and commercial use. There has been a rapid increase in the use of wind energy worldwide.

4.2 The Nature of Wind

Wind is a form of solar energy caused by a combination of three concurrent events:

1. The sun unevenly heating the atmosphere
2. Irregularities of the earth's surface
3. The rotation of the earth.

Local winds

Land and water are heated differently. At daytime, solar radiation readily converts to sensible energy on the land surface. However, it is partly absorbed in the layer below the water surface and partially evaporates as some of the water evaporates. Air above the land gets warmer because of the higher temperature of the land mass than the air above water. Warmer lighter air above the land rises, and cooler heavier air above the water replaces it. This is the mechanism of shore breezes. At night, the direction o the breezes is reversed because the land mass cools to the sky. Hills and mountain sides are responsible for local winds. Summer temperatures are higher above slopes, while nighttime temperatures are lower. The result is that heated air rises along the slopes during the day and relatively cool heavy air flows down at night as a result.

Global Winds

There are global winds that occur in belts that travel around the globe. In the same way as local winds, global winds are produced by unequal heating of the atmosphere. At the equator, the temperature is highest, and at the poles, it is coldest. Due to the differences in heating, the troposphere experiences massive convection currents. As warm air rises, it reaches the tropopause at the equator. The current flows north or south when it is unable to rise any higher. A cooling process takes place when moving air reaches towards the poles. It sinks to the surface when it gets cool. Then, it flows over the surface and back towards the equator.

A particular site experiences remarkable consistency in wind patterns every year, despite the wind's intermittent nature. Wind speeds are higher in hilly and coastal areas than they are inland.

The basic energy conversion process is as follows:

1. Solar energy is converted to wind energy.
2. Wind energy is converted to motion energy by the blades, which rotates the shaft.
3. The rotational velocity of the shaft is increased through the gear box. This does not increase the energy, and in fact, loses some due to frictional heat loss in the gear box.
4. The gear box spins another shaft, which spins the generator.
5. The generator generates an electrical current.
6. Wind turbines convert wind to electricity. This is different than a windmill, which converts wind energy into mechanical energy that is used for another purpose (e.g., pumping water or in the "old days," grinding grain).
7. A wind farm is a number of turbines that are installed in the same area, usually by the same company.

4.3 Potential of Wind Energy in India

Wind is an intermittent and site-specific resource of energy and therefore, an extensive Wind Resource Assessment is essential for the selection of potential sites. The Government, through National Institute of Wind Energy (NIWE), has installed over 800 wind-monitoring stations all over country and issued wind potential maps at 50 m, 80 m, 100 m and 120 m above ground level. The recent assessment indicates a gross wind power potential of 302 GW in the country at 100 meter and 695.50 GW at 120 meter above ground level. Most of this potential exists in seven windy states as given below Table –

Table 4.1: Wind energy potential in different states of India

S.No.	State	Wind Energy Potential at 100 m (GW)	Wind Energy Potential at 120 m (GW)
1	Gujarat	84.43	142.56
2	Rajasthan	18.77	127.75
3	Maharashtra	45.39	98.21
4	Tamil Nadu	33.79	68.75
5	Madhya Pradesh	10.48	15.40
6	Karnataka	55.85	124.15
7	Andhra Pradesh	44.22	74.90
8	Total 7 windy states	292.97	651.72
9	Others	9.28	43.78
	Total	302.25	695.50

4.4 Relationship Between Wind Speed and Height

Generally, wind speed varies with height which can be seen clearly in Figure 4.1. The wind speed increases as the distance above the ground increases under normal conditions. As the height increases, turbulence and surface features are less significant. In addition to distance from the ground, roughness of the terrain affects variability. A power law represents the vertical variation of wind speed expressed as follows.

$$\frac{v}{v_0} = \left(\frac{h}{h_0}\right)^{\alpha}$$

Where, v is the wind speed at the required height h, v_0 is wind speed at the original height h_0, and α is the surface roughness coefficient. The value of the exponent α varies from less than 0.10 over the tops of steep hills to over 0.25 in sheltered locations. The typical value for flat coastal regions is 0.143, which is termed the one-seventh power law. From measurements at different heights the exponent can be determined.

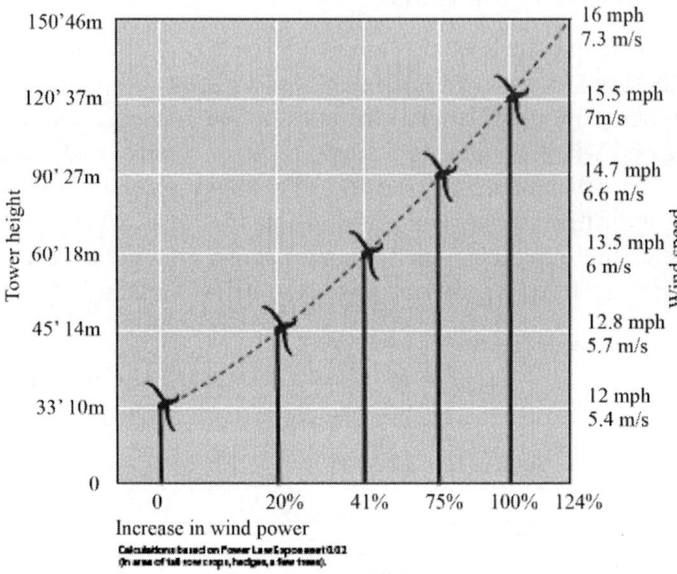

Fig. 4.1: Relation between the wind speed and height

4.5 The Power in Wind

Wind is a source of energy due to its motion. The energy in moving air can be extracted and converted into useful work by any device that can slow it down, such as a sail or propeller. In order to determine the output power generated by a wind mill, there are three factors to consider:

a. The wind speed
b. The cross section of wind swept by rotor, and
c. The overall conversion efficiency of rotor, transmission system and generator or pump

It is impossible, no matter how well designed, for a device to extract all of the wind's energy because the wind would have to be slowed, thus preventing more air from passing through the rotor. Approximately one-third of the free velocity of the intercepted air can be reduced by the rotor. It is therefore estimated that a 100% efficient aerogenerator could only convert around 60% of wind energy into mechanical energy.

The power generated by the wind can be computed using Kinetics (Kinetic means relating to motion or resulting from motion). Wind mills convert wind energy into mechanical energy by converting its Kinetic energy. In general, power is defined as energy per unit time. In this case, the available energy is the kinetic energy of the wind. The kinetic energy of any particle is equal to one half its mass times the square of its velocity.

Kinetic Energy of particle = ½ mv^2

Where, m= mass of particle (kg), v : velocity of wind (m/s)

The amount of air passing in unit time, through an area 'A', with velocity 'V' is A x V, and its mass 'm' is equal to its volume multiplied by its density 'ρ' of air.

$m = \rho\, AV$

Where, m is the mass of air transversing the area 'A' swept by the rotating blades of a wind mill type generator.

$$\text{Kinetic energy} = \frac{1}{2} AV \times V^2$$

$$-\frac{1}{2} AV^3 \text{ watts}$$

4.6 Power Coefficient

The fraction of the free-flow wind power that can be extracted by the rotor is called the power co-efficient; Thus,

Where, power available is calculated from the air density, rotor diameter and free wind speed as discussed earlier. The maximum theoretical power coefficient is equal to 16/27 or 0.593. This value cannot be exceeded by a rotor in free-flow wind-stream beyond 0.593. But such a rotor would not be strong enough to withstand the stresses to which it is subjected when rotating at a high rate in a high-speed wind stream.

$$\text{Power Coefficient}\left(Cp\right) = \frac{\text{Power of wind rotor}}{\text{Power available in wind}}$$

The German physicist Albert Betz concluded, in 1919, that a wind turbine can convert no more than 16/27 (59.3%) of air kinetic energy into mechanical energy. It is known as Betz's Limit or Betz's Law. This limit is not due to the inefficient generator but depends on the nature of the wind turbines themselves. The wind is slowed down by wind turbines in order to extract energy from it. To be 100% efficient, a wind turbine must stop 100% of the wind - however, the rotor would then not turn and there wouldn't be any conversion of kinetic energy. Alternatively, if a wind turbine had just one blade, most wind passing through the area covered by the blade would miss the blade entirely. This would result in most of the kinetic energy remaining with the wind and not being transmitted to the wind turbine's generator.

4.7 Cut-in and Cut-out Speed

As soon as the blades start spinning and generating power (typically between 10-15 km/h), it is called cut-in speed. A wind turbine generates more electricity as its speed increases until it reaches its rated speed. It is at this point that the turbine produces its maximum power, or its rated power. Power is generated by the turbine as wind speed continues to increase until eventually it hits a cut-out speed (varies by turbine) and shuts off to prevent unnecessary strain on the rotor. Figure 4.2 shows the different stages of power generation of the wind turbine.

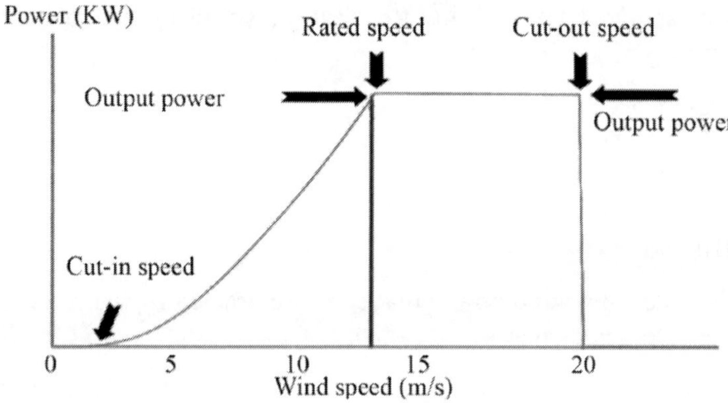

Fig. 4.2: Cut in and cut out speed

4.8 Maximum power in wind

It is impossible to convert all of the wind's energy into mechanical energy. As an example, consider a horizontal-axis, propeller-type windmill, which is the most common type today. Consider the turbine wheel with thickness 'ab' as shown in Figure 4.3.

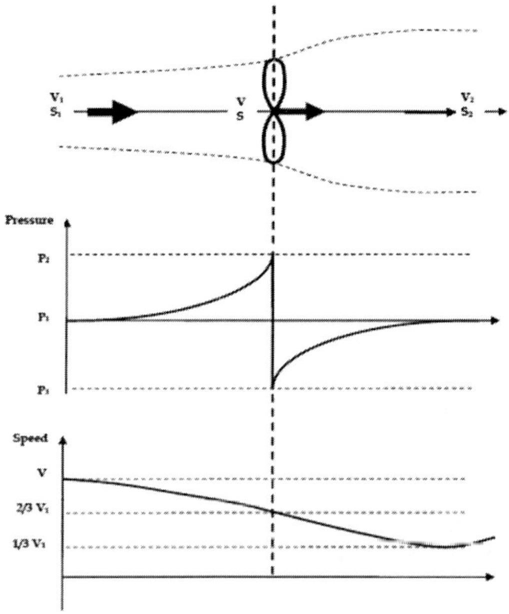

Fig. 4.3: Profile of wind during power generation

Let

Pi = wind pressure at upstream of turbine

Vi = velocity at the upstreme of turbine

Pe = pressure at down streme of the turbine

Ve = velocity at the downstreme of the turbine

Ve is less than Vi, because kinetic energy extracted by the turbine

For optimum exit velocity 'Ve,opt', that results in maximum power 'Pmax', which can be obtained by differentiating 'P' with respect to 'Ve', and equating the derivative to zero, i.e.,

$$P_iV + \frac{V_i^2}{2g_c} = P_aV + \frac{V_a^2}{2g_c}$$

$$\frac{dP}{dV_e} = \frac{1}{4g_c}\rho A\left[(1+0)\left(V_i^2 - V_e^2\right) + \left(V_i + V_e\right)\left(-2V_e\right)\right] = 0$$

$$V_i^2 - V_e^2 - 2V_eV_i - 2V_e^2 = 0$$

$$V_i^2 - 2V_eV_i - 3V_e^2 = 0$$

$$3V_e^2 + 2V_eV_i - V_i^2 = 0$$

This is solved for positive Ve to give Ve.opt. The quadratic equation has two solutions, i.e Ve =Vi and Ve=1/3 Vi, and only second solution is acceptable, Thus

$$V_{e.opt} = \frac{1}{3} V_i$$

$$P_{max} = \frac{8}{27g_c} \rho A V_i^3$$

$$= \frac{16}{27g_c} \left(\frac{1}{2} \rho A V_i^3 \right)$$

$$= 0.593 \times \frac{1}{2} \frac{\rho V_i^3 A}{g_c}$$

$$=0.593P_{total}$$

A wind turbine's maximum theoretical efficiency, also known as the power coefficient (Cp), is the ratio of output power to total wind power. The factor 0.593 is known as the Betz coefficient. It is the maximum fraction of the power in a wind stream that can be extracted.

4.9 Forces on the Blades and Thrust on Turbines

there are two types of forces which are acting on the blades of propeller-type wind turbine. They are

a) Circumferential force acting in the direction of wheel rotation that provides the torque

b) Axial force acting in the wind stream that provides an axial thrust that must be countered by proper mechanical design.

The circumferential Force, or Torque (T) can be obtained from

$$T = \frac{P}{\omega} = \frac{P}{\pi D N}$$

Where,

T= Torque, kgf or Newton

ω = angular velocity of turbine wheel, m/s

D = Turbine diameter, m

$$P = \eta \frac{1}{2g} \rho A V^3$$

$$P = \eta \frac{1}{2g} \rho A V^3$$

Vertical-axis turbines

A vertical-axis wind turbine can be classified into several varieties, including the eggbeater-type Darrieus model. Due to their omnidirectional design, these turbines don't need to be adjusted to point into the wind. Different types of turbines can be seen in Figure 4.5.

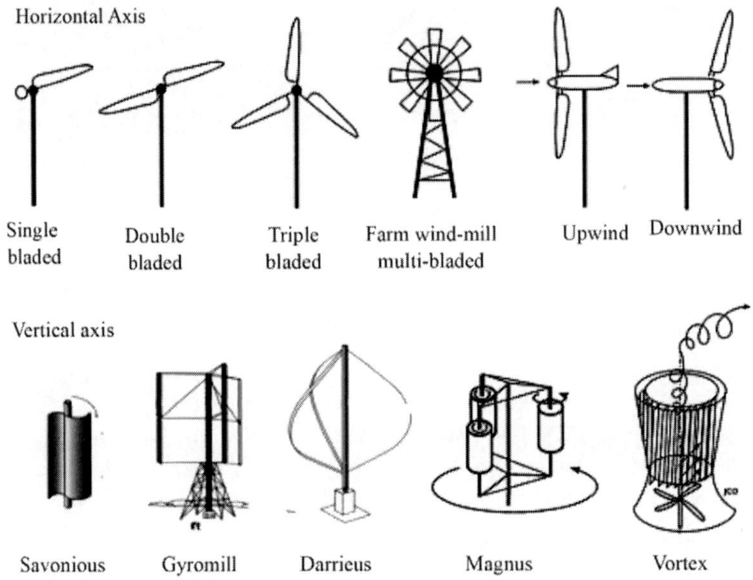

Fig. 4.5: Types of Horizontal and vertical axis wind turbine

4.11 Horizontal Axis Wind Turbines (HAWT)

The horizontal axis wind turbine (HAWT) is the most common design of wind turbines. As shown in figure, a HAWT has blades that look like propellers that spin on a horizontal axis like a windmill. Rotor shafts and electrical generators on horizontal axis wind turbines are mounted on top of a tower, facing the wind. Turbines with small rotors (blades) are pointed by a simple wind vane, while larger turbines use a wind sensor and servo motor to turn into the wind. There is usually a gearbox on large wind turbines that converts the slow rotation of the rotor into a faster rotation that can drive an electrical generator. Since a tower produces turbulence behind it, the turbine is usually pointed upwind of the tower. To prevent wind turbine blades from being pushed into the tower by high winds, wind turbine blades are made stiff. A small amount of tilt is sometimes applied to the blades, and the blades are placed a considerable distance from the tower. Despite the problem of turbulence, downwind machines have been built because they do not require additional mechanisms to stay aligned with the wind. Furthermore, in high wind conditions, the blades can bend, reducing their swept area and wind resistance. Most HAWTs are upwind machines because turbulence which causes fatigue failures.

Important point to remember recording HAWT:

a) Lift is the main force

b) Much lower cyclic stress

c) 95% of the existing turbines are HAWTs

d) Nacelle is placed at the top of the tower

e) Yaw mechanism is required

HAWT Advantages

a) A tall tower base allows stronger wind to be harnessed at sites with wind shear. At some wind shear sites, wind speed increases by 20% every ten meters up, while power output increases by 34%

b) It has a high efficiency due to its blades always moving perpendicular to the wind, receiving power throughout the whole rotation

HAWT Disadvantages

a) Huge towers need to be constructed because blades, gearbox, and generator are heavy

b) Components of horizontal axis wind turbine (gearbox, rotor shaft and brake assembly) being lifted into position

c) Due to their height, they disrupt the landscape appearance and sometimes cause local opposition

d) The majority of HAWTs use an upwind design, with the rotor facing the wind in front of the tower, since download variants suffer from fatigue and structural failure caused by turbulence

e) HAWTs are generally required to have a braking or yawing device in high winds in order to prevent the turbine from spinning and destroying or damaging itself.

4.12 Components of Wind Mill

The simple structure of horizontal axis wind turbine (wind mill) shown in Figure 4.6.

The following components are used in a wind mill.

a. **Anemometer:** Measure the wind speed and transmits wind speed data to the controller

b. **Blades:** Most turbines have either two or three blades. Wind blowing over the blades causes the blades to "lift" and rotate.

c. **Brake:** A disc brake, which can be applied mechanically, electrically, or hydraulically to stop the rotor in emergencies.

d. **Controller:** The controller starts up the machine at wind speeds of about 8 to 16 miles per hour (mph) and shuts off the machine at about 55 mph. Turbines do not operate at wind speeds above about 55 mph because they might be damaged by the high winds.

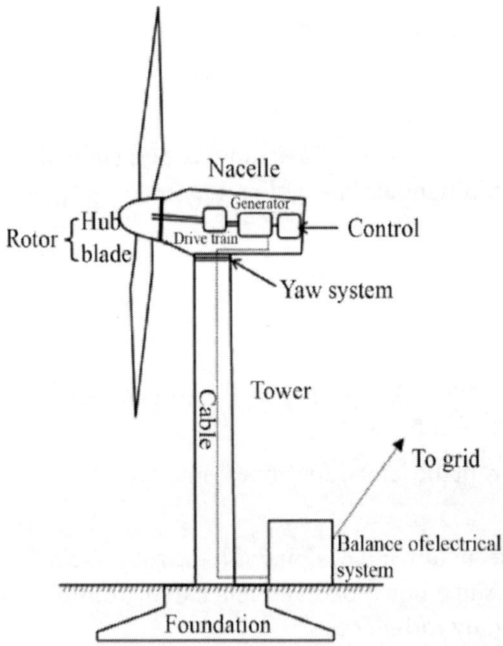

Fig. 4.6: Components of horizontal axis wind turbine

e. **Gear box:** Gear connect the low-speed shaft to the high speed shaft and increase rotational speeds from about 30 to 60 rotations per minute (rpm) to about 1000 to 1800 rpm, the rotational speed required by most generators to produce electricity. The gear box is a costly (and heavy) part of the wind turbine and engineers are exploting "direct-drive" generators that operate at lower rotational speeds and don't need gear boxes.

f. **Generator:** Usually an off-the-shelf induction generator that produces 60-cyclic AC electricity.

g. **High speed shaft:** Drives the generator.

h. **Low-speed shaft:** The rotor turns the low-speed shaft 30 to 60 rotations per minute.

i. **Nacelle**: The nacelle sits the lower and contains the gear box, low-and high-speed shafts, generator, controller, and brake. Some nacelle are large enough for a helicopter to land on.

j. **Pitch**: Blades are turned, or pitched, out of the wind to control the rotor speed and keep the rotor form turning in winds that are too high or too.

k. **Rotor**: The blades and the hub together are called the rotor

l. **Tower**: Towers are made from tubular steel (shown here), concrete, or steel lattice. Because wind speed increases with height, taller towers enable turbines to capture more energy and generate more electricity.

m. **Wind direction**: This is an "upwind" turbine, so called because it operates facing into the wind. Other turbines are designed to run "downwind", facing away from the wind

n. **Wind vane**: Measure wind direction and communicates with the yaw drive to orient the turbine properly with respect to the wind

o. **Yaw drive**: Upwind turbines face into the wind, the yaw drive is used to keep the rotor facing into the wind as the wind direction changes. Downwind turbines don't require a yaw drive; the wind blows the rotor downwind.

p. **Yaw motor**: Powers the yaw drive.

4.13 Vertical Axis Wind Turbines(VAWT)

The main rotor shaft of vertical wind turbines (VAWT) is arranged vertically (Figure 4.7). With this arrangement, wind turbines are not required to be faced in the direction of the wind. An advantage of using VAWT is when the direction of the wind changes rapidly or if there is turbulence. It is easier to maintain a vertical axis, since the generator and other primary components do not require support from the tower and can be placed near the ground. VAWTs have the main disadvantage of creating drag when rotating into the wind. As wind speed decreases at a lower altitude, less wind energy is available for a given turbine size. Air flowing near the ground and other objects can cause turbulent flow, which can cause vibration, noise and bearing wear that may increase maintenance or shorten its service life. On the other hand, turbines mounted on rooftops are generally redirected over the roof by the building, which can double the wind speed. In order to maximize wind energy and minimize wind turbulence, the rooftop mounted turbine tower should be approximately 50% of the building height.

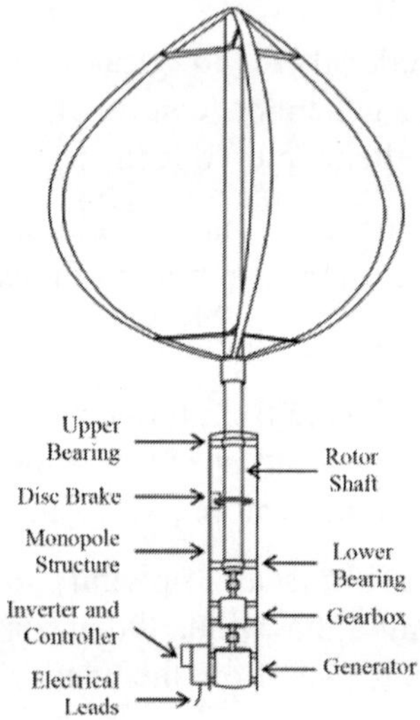

Fig. 4.7: Component of vertical axis wind turbine

Important points about VAWT

- Nacelle is placed at the bottom.
- Drag is the main force
- Yaw mechanism is not required
- Lower starting torque
- Difficulty in mounting the turbine
- Unwanted fluctuations in the power output

Advantages of VAWT

a) No yaw mechanisms is needed

b) A VAWT can be located nearer the ground, making it easier to maintain the moving parts.

c) VAWTs have lower wind startup speeds than the typical the HAWTs.

d) VAWTs may be built at locations where taller structures are prohibited.

e) VAWTs situated close to the ground can take advantage of locations where rooftops, means hilltops, ridgelines, and passes funnel the wind and increase wind velocity.

VAWT Disadvantages

a) The VAWT is generally less efficient than a HAWT due to the additional drag caused by the blades rotating into the wind.

b) Wind speeds are lower at ground level, so rotors are placed closer to the surface, thus not utilizing higher winds.

c) Due to the disadvantages mentioned above, VAWTs do not commonly operate in the wind industry

4.14 VAWT Subtypes

Darrieus wind turbine

This windmill is often referred to as an "eggbeater windmill" because of its long, thin blades in the shape of loops attached to the axle as shown in Figure 4.7. The design was invested in 1931 by French engineer Georges Darrieus. Due to their resemblance to a giant eggbeater, Darrieus wind turbines are commonly called "Eggbeater" turbines. They are efficient, but their towers suffer from cyclic stress and torque ripple, which makes them unreliable. A disadvantage of the Darrieus wind turbines is that they cannot start themselves. A rotor is generally required to start them, as the starting torque is very low, and they require external power. A rotor with three or more blades has a higher solidity, which reduces torque ripples. These turbines are generally supported by using guy-wires. Newer Darrieus-type turbines have an external superstructure connected to the top bearing. A tip speed ratio (TSR) measures the difference between the rotating velocity of the turbines and the wind velocity. In the Darrius wind turbine, the TSR is greater than 1, meaning the rotational speed here is greater than the wind speed and creates less torque. As a result, Darrieus turbines make excellent generators of electricity. As a result of the centrifugal forces generated during rotation, the turbine blades must be strengthened, while the generator itself can withstand less force than the Savorius.

Advantages

a. Rotor shafts are vertical. This allows the load, such as a generator and centrifugal pump to be placed below ground level.

b. It is possible to take wind from every direction on the rotor.

c. A vertical axis windmill might have a better visual appeal than a horizontal axis windmill.

d. Can be easily integrated into buildings.

Disadvantages

a. Difficult start unlike the Savonius wind turbine

b. Low efficiency.

Savonius wind turbine

A Savonius wind turbine (Figure 4.8) is a vertical-axis wind turbine invented by the Finnish engineer Sigurd Savonius in the 1920's. The design of this wind turbine is one of the simplest. A wind turbine is driven by two to three "scoops" that use drag action to convert wind energy into torque. An S-shaped turbine looks like a two-scoop Savonius when viewed from above in cross-section. Due to the curvature of the scoops, the turbine experiences less drag when moving against the wind than with it, and this causes the turbine to spin regardless of the direction of the wind. Savonius turbines and other drag-type turbines are less efficient in using the wind's energy than lift-type turbines, which are common in wind farms. Savonius turbines are drag type turbines that are commonly used in ventilation and anemometers where eliability is important. Due to their drag type, they are less efficient than the common HAWT. Savonius excels at self-starting and turbulent wind areas.

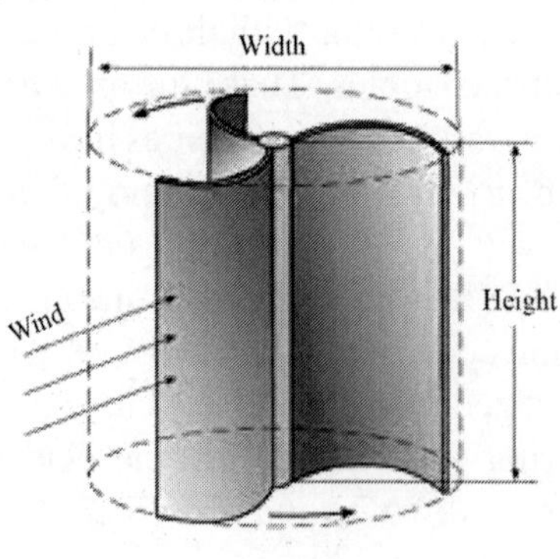

Fig. 4.8: Savonius wind turbine

Advantages

a) Since it rotates on a vertical axis, the Savonius turbine remains effective even when the wind direction changes

b) Since the Savonius design works well even at low wind speeds, it doesn't require a tower or other expensive structure to hold it in place which reduces overall costs.

c) There is no sound coming from the device, it's easy to assemble, and it's relatively small.

d) Maintenance is easy since the turbine is close to the ground.

Disadvantages

Compared to a conventional turbine, the scoop system captures half the amount of energy from the wind.

4.15 Wind Farm

Wind farms are sites designed for the generation of wind power. The wind generators range from 600 kW to 5 MW and are spread across the land to be able to use the wind as efficiently as possible.

Wind farms can be classified into three types.

a. On-shore: The most common type of wind farm is located "on-shore" (at least 3 km inland)

b. Near-shore: these farms are less than 3 km from the coast

c. Off-shore: Off-shore (in the open sea or in lakes many miles from the shore) and are located in places where they do not infringe upon pre-existing businesses or important sailing routes.

Working of wind farm

Each wind turbine (Figure 4.9) produces electrical power that is channelled downward into electrical ground cables. Ground cables are also used to transmit signals that control wind turbine functions. Wind farms have underground cable ducts connecting the turbines together. Each wind farm has a delivery cabin-station. Afterward, the electrical power produced is channelled into the electric grid, to the utility site. All wind farms (Figure 4.10) are equipped with step-up substations to collect the energy generated by the turbines and transmit it through the medium voltage cables. Normally a substation will be composed by the following elements:

- Medium voltage system
- High voltage system
- Capacitors banks

- Auxiliary services
- Control, protection and metering system
- Communication system
- Fire protection and intruders protection systems

A medium voltage system consists of a busbar, disconnectors, circuit breakers and current transformer. In a high-voltage system, transformers, surge arresters, current transformers, voltage transformers, circuit breakers, and disconnectors with earthing are used. An electric system with a high voltage usually includes a transformer as well as circuit breakers, surge arresters, current transformers, voltage transformers, and earthing reactances. Active and reactive power requirements are met through the installation of capacitor banks. The substation has auxiliary services such as AC and DC energy, as well as a large battery bank that can provide power for several hours in case of an emergency, a rectifier, and a diesel generator for backup power. The wind farm must be operated in accordance with local regulations and grid requirements. For the substation to operate correctly, communication between the wind farm substation, adjacent substations, and the grid control center is essential. As for the intruders' protection system, it usually consists of a fence and a closed circuit TV as well as optical or infrared detectors.

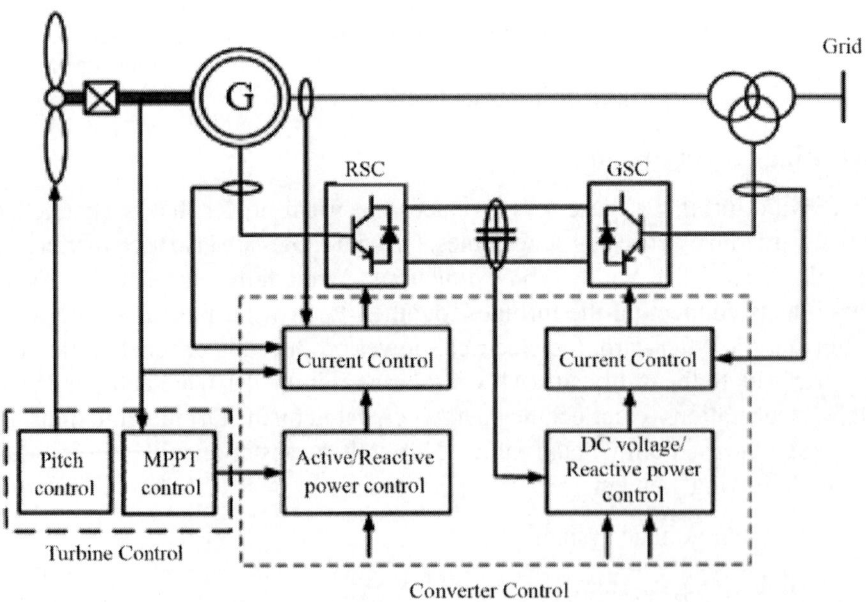

Fig. 4.9: Single wind turbine grid connection

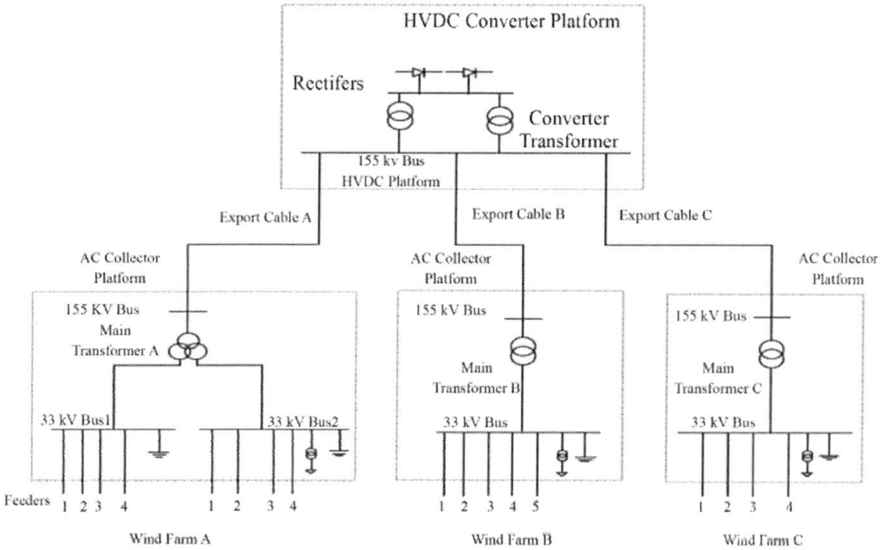

Fig. 4.10: Wind farm working

4.16 Site Selection for Wind Farm

It is most advantageous to locate wind energy conversion machines where the wind is strong and persistent, since wind power increases rapidly with speed. A given site's daily winds may vary greatly, but its monthly and also annual averages are remarkable constants. Winds with speeds above average are the main contributors to the wind power available at a particular site. Wind turbines are most suitable for sites with moderately high or high annual average wind speeds. Aero turbine generators feed power into a conventional electric grid, which is why choosing the right location for a wind energy conversion system (WECS) is essential. The net wind electric energy generated each year may be lower if the WECS sites are poorly chosen. This may result in high capital costs for the equipment, high prices for wind generated electricity, and low returns on investment. Regardless of whether the WECS is a small generator not tethered to the grid, the site must be carefully selected to avoid excessive break even periods. Prior to erecting a wind turbine/ wind farm at a specific site, technical, economic, environmental, social, and other actors are taken into account.

Some of the main site selection consideration are given below:

1. **High annual average wind speed:** The power generated by the wind mill is directly proportional to the wind velocity, so small increases in wind velocity have a significant impact on its power generation.

For example, Doubling the velocity, increases power by a factor of 8. High wind velocity sites are clearly preferred for WECS applications. Therefore, a high average wind velocity is a fundamental parameter to take into consideration when assessing a WESCS site.

2. **Availability of anemometry data:** Before a sitting decision is made, the aenometry data should be available at the precise location where any proposed WECS will be built.

3. **Availability of wind V(t) Curve at the proposed site:** This important curve determines the maximum wind energy and therefore is the primary controlling factor for predicting the electrical output and revenue return of WECS machines. It is desirable to have average wind speed 'V' such that V $\geq$12-16 km/hr (3.5 – 4.5 m/sec) which is about the lower limit at which present large scale WECS generators 'cut in' i.e., start turning. It is also the V(t) Curve that determines the reliability of the generator power delivered by the WECS for, if it reaches zero, there will be no power generated. Short calm periods will lower WECS reliability than long calm periods. In order to make accurate reliability estimates, it is best to measure V(t) Curve over a period of approximately 5 years.

4. **Wind structure at the proposed site:** An ideal site for the WECS would have a flat V(t) Curve, i.e., always blowing wind. However, most sites are far from ideal. Turbulent and gusty wind near the ground changes rapidly in direction and velocity.

5. Altitude of the proposed site: The density of the air affects the power in the wind, and thus the useful electric power output of the WECS. It is also well known that at higher altitudes the wind tends to have higher velocities. The distinction between altitude and height above ground must be made carefully. WECS sites at sea level are different from those at other elevations.

6. **Terrain and its aerodynamic:** It is critical to understand the terrain of the site to be chosen. Wind velocity can be increased by placing the WECS near the top, but not on top of a hill that faces the prevailing wind. Additionally, the wind may not flow horizontally here, so the rotor must be tipped so that it is always perpendicular to the actual wind flow. The prevailing wind can be channelled into a pass region by hills or mountains, thereby generating higher wind power.

7. **Local Ecology:** When the surface is base rock, the hub height may be lower, resulting in a lower structure cost. In the presence of trees, grass, or vegetation, which tend to destructure the wind, higher hub heights are required, resulting in higher system costs.

8. **Distance to road or railways:** Heavy machinery, structures, materials, blades, and other apparatus will need to be moved into any WECS site chosen by the system engineer.
9. **Nearness of site to local centre/users:** Transmission line length can be minimized to lower the losses and costs. The proposed WECS sites should be relatively close to the users of the generated electricity.
10. **Nature of ground:** In order for a WECS to be successful, the ground conditions must be such that the foundation can be securely anchored. It is important that the ground surface is stable. A WECS foundation should not be subjected to erosion, as it could wash out later, destroying the entire system.
11. **Favourable land cost:** In addition to other siting costs, land costs should be favorable as they are included in total WECS system costs.
12. The site should be free of other conditions such as icing, salt spray, and blowing dust, since they may affect the blades of the turbines or negatively impact machinery and electrical equipment.

Multiple Choice Questions

1.Which of the following is the energy used for storing Wind energy?

a) Kinetic b) Potential
c) Chemical d) Electrical

2.Which of the following are renewable energies?

a) Wind b) Water
c) Wood d) All the above

3.Which of the following are the main components used during generation of wind type energy?

a) Fan b) Wind
c) Batteries d) All the above

4.Which of the following are the events that allows in generation of electricity?

a) Solar heat b) Earth surface
c) Moving of earth d) All the above

5. Which of the following is the purpose of wind energy generation?

a) Grinding grain b) Pumping water

c) Generating power d) All the above

6. Which of the following are the basic components of a horizontal type wind turbines?

a) Gear box b) Controller

c) Accelerometer d) All the above

7. Which of the following component is connected to gear box in horizontal type wind turbine?

a) Low speed shaft b) High speed shaft

c) Controllers d) Accelerometer

8. Which of the following is the function of a low speed type shaft in wind turbine?

a) Speed of shaft is increased

b) Speed of shaft is minimized

c) Speed of shaft is maintained moderate

d) Speed of shaft is maintained at zero

9. Which of the following component is connected to generator box in horizontal type wind turbine?

a) Low speed shaft b) High speed shaft

c) Controllers d) Accelerometer

10. A high speed type shaft in horizontal wind mill translates ________ to AC type voltage.

a) Rotatory motion b) Swing motion

c) Linear motion d) None of the above

11. What happens to speed when in high speed type shaft of horizontal wind turbine?

a) Zero b) Critical

c) Infinity d) One

12. Which of the following component does not exist in direct driven type horizontal wind turbine?

a) Gear box b) Blades

c) Controllers d) Anemometer

13. Which of the following is the major component of a commercial type wind turbines?

a) Computer b) PLC's

c) Shafts d) Both a and b

14. Which of the following are the three ways to control speed of wind turbines?

a) Adjusting yaw and pitch

b) Allowing turbine to run at any speed bounded with ratings

c) Controlling of generators frequency output for double feed

d) All the above

15. A controller in commercial type wind turbines are provided with data from _______.

a) Anemometer b) Yaw drive

c) Rotor d) Wind vane

16. Which of the following phases are used in vertical type wind turbine?

a) Single b) Two

c) Three d) Both a and c

17. Which of the following is used for filtering out input frequency and converting in to DC in vertical type wind turbine?

a) Rectifier b) Passive filter

c) Active filter d) Both a and b

18. Which of the following are components of power electronic type frequency converter used in vertical type wind turbines?

a) Rectifier b) Inverter

c) Transformer d) All the above

19. What is the height of HAWT tower?

a) 40 to 100 meters b) 50 meters

c) 110 meters d) 30 meters

20. Which of the following are the components of horizontal type wind turbine?

a) Break b) Gears

c) AC generators d) All the above

21. How many number of blades allow wind turbine to operate at higher speed?

a) More b) Lesser

c) Zero d) Single

22. Which of the following are the advantages of three blade turbine?

a) Does not makes sounds much during operation

b) Less vibration

c) Lowest cost of energy

d) All the above

23. Which of the following are the disadvantages of three blade type turbine?

a) Higher in weight b) Expensive

c) Employs yaw control d) All the above

24. Which of the following country stands high in generation of wind power?

a) China b) India

c) US d) Romania

25. Which of the following is the advantage of wind energy?

a) No carbon dioxide emission b) Eco-friendly

c) More power generation d) All the above

26. A windmill's blades are pushed outward by wind force in which of the following operations?

a) Drag design b) Lift design

c) Both a and b d) None of the above

27. When a gearbox is added to a wind mill, what happens to its speed?

a) Increases b) Decreases

c) Constant d) Zero

28. What is the minimum wind speed required for blades to rotate and produce power?

a) Cut in speed b) Rated speed

c) Cut out speed d) Betz limit

29. What is the minimum speed at which a windmill produces the required power rate?

a) Cut in speed b) Rated speed

c) Cut out speed d) Betz limit

30. In windmills, what is the speed at which they cease to function?

a) Cut out speed b) Betz limit

c) Rated speed d) Cut in speed

31. The quantity of air passing over blades and through the rotor of a wind turbine is defined as one of the following characteristics?

a) Rated speed b) Cut out speed

c) Cut in speed d) Betz limit

32. How much wind is theoretically collected by wind turbines?

a) 40% b) 50%

c) 59% d) 80%

33. Wind energy is a ____ source of energy.

a) Intermittent b) Portable

c) Constant power d) None of the above

34. What factors should an engineer consider when designing a windmill?

a) Direction of wind b) Height of tower

c) Climatic condition d) All the above

35.In a wind turbine, if velocity of wind is doubled, within the opeating rang, the power output will be

a) Six times
b) Eight times
c) Doubled
d) Reduce to half

Answer

1	a	2	a	3	d	4	d	5	d
6	d	7	a	8	a	9	b	10	a
11	b	12	a	13	d	14	d	15	a
16	d	17	d	18	d	19	a	20	d
21	b	22	d	23	d	24	a	25	d
26	c	27	a	28	a	29	b	30	a
31	d	32	c	33	a	34	d	35	b

5

Biomass Energy

5.1 Introduction

Biomass energy is the energy obtained from biomass or living things. Biomass is renewable organic material that comes from plants and animals. Plants generally produced through the process of photosynthesis as shown in Figure 5.1. Biomass is organic, meaning it is made of material that comes from living things, such as plants and animals. Wood, plants, and waste are the most common biomass energy sources. These are called biomass feedstocks. A plant absorbs the sun's energy through photosynthesis and converts carbon dioxide and water into nutrients (carbohydrates). The energy from these materials can be transformed into usable energy through direct and indirect means. Biomass can be burned to create heat (direct), converted into electricity (direct), or processed into biofuel (indirect).

The use of biomass remains important in many countries, especially in developing countries for cooking and heating. As a means of reducing carbon dioxide emissions from fossil fuel usage, biomass fuels are increasingly used for transportation and electric generation in developed countries. A biomass is a variety of materials that include wood and agricultural residues as well as animal and human waste.

- Wood and wood processing wastes—firewood, wood pellets, and wood chips, lumber and furniture mill sawdust and waste, and black liquor from pulp and paper mills
- Agricultural crops and waste materials—corn, soybeans, sugar cane, switchgrass, woody plants, and algae, and crop and food processing residues, mostly to produce biofuels
- Biogenic materials in municipal solid waste—paper, cotton, and wool products, and food, yard, and wood wastes
- Animal manure and human sewage for producing biogas/renewable natural gas

In the process of photosynthesis, plants convert radiant energy from the sun into chemical energy in the form of giucose-or sugar

(water) (carbon dioxide) (sunlight) (glucose) (oxygen)

$6H_2O + 6CO_2 + \text{radiant energy} \longrightarrow C6H_{12}O_6 + 6O_2$

Fig. 5.1: Process of photosynthesis

5.2 Biomass to Energy Conversion

Biomass is converted to energy through various processes, including:

- Direct combustion (burning) to produce heat
- Thermochemical conversion to produce solid, gaseous, and liquid fuels
- Chemical conversion to produce liquid fuels
- Biological conversion to produce liquid and gaseous fuels

In general, biomass is converted to energy through direct combustion. There are many uses for biomass, including heating buildings and water, using it for energy production in the industrial sector, and using it to generate electricity through steam turbines.

A thermochemical process for converting biomass includes pyrolysis and gasification. A biomass feedstock material is heated at high temperatures in closed, pressurized vessels called gassifiers during thermal decomposition. The main difference between the two is the temperature and amount of oxygen present during the conversion process.

- A pyrolysis process involves heating organic materials to 400–500 °C in the absence of free oxygen. As a result of biomass pyrolysis, charcoal, bio-oil, renewable diesel, methane, and hydrogen can be produced.
- Hydrotreating is used for producing bio-oil (produced by fast pyrolysis) which can be converted into renewable diesel, renewable gasoline, and renewable jet fuel by combining it with hydrogen under elevated temperatures and pressures.

- In gasification, organic materials are heated to 800–900 °C and then oxygen and/or steam are injected into a vessel to produce synthesis gas, or syngas, which is rich in carbon monoxide and hydrogen. Syngas can be used in diesel engines, to generate electricity in gas turbines, and for heating. The hydrogen can also be separated from the gas, and the hydrogen can be burned or utilized in fuel cells. By using the Fischer-Tropsch process, liquid fuels can be produced from syngas.

Transesterification converts vegetable oils, animal fats, and greases into fatty acid methyl esters (FAME), which are used to make biodiesel. Biological conversion includes fermentation to convert biomass into ethanol and anaerobic digestion to produce renewable natural gas. Ethanol is used as a vehicle fuel. Biogas and biomethane, also called renewable natural gas or biomethane, are produced in anaerobic digesters in sewage treatment plants and at dairy and livestock operations. Additionally, it can be captured from solid waste landfills. Renewable natural gas can be used in the same way as fossil fuel natural gas when it is properly treated.

5.3 Biomass Combustion System

Biomass can be converted to heat most efficiently using direct combustion, which has been around for many years. Combustion of biomass fuel produces heat when excess air is burned. Combustible vapours are formed from the biomass at the first stage of combustion, which burn as flames. a forced air supply is used to burn the residual material as charcoal, which provides more heat. Sometimes the hot combustion gases are directly used for product drying, but they are more commonly used to generate hot air, hot water, or steam through a heat exchanger.

In order to achieve good combustion efficiency, oxygen in the air must be in good contact with biomass fuel. Biomass combustion results in the release of carbon dioxide and water vapor, but also tars, smoke, and alkaline ash particles. In designing environmentally acceptable biomass combustion systems, it is important to minimize these emissions and accommodate their potential effects.

There are several furnace designs that can be used to burn biomass, producing hot gases, hot air, hot water, or steam, capturing 65-90% of the fuel's energy. Fuels with higher moisture content generally have lower efficiencies. Typical capital costs for biomass power projects range from Rs 3 crores/ MW to Rs 4 crores/MW. Costs of generation depend upon the cost of biomass, the plant load factor, and the efficiencies of conversion. There are a variety of combustion furnaces or combustors available worldwide that are designed to accommodate a variety of fuel characteristics and combustion requirements

Underfeed Stokers

Biomass is fed into the combustion zone from underneath a firing grate as shown in Figure 5.2. Small-scale systems with boilers up to 6 MWth can be equipped with these stoker designs, as long as these biomass fuels have low ash contents. It is necessary to use more efficient ash removal systems for fuels that contain high levels of ash, such as bark, straw, and cereals. If molten or sintered ash particles cover the upper surface of the fuel bed, they may cause problems in underfeed stokers due to unstable combustion conditions.

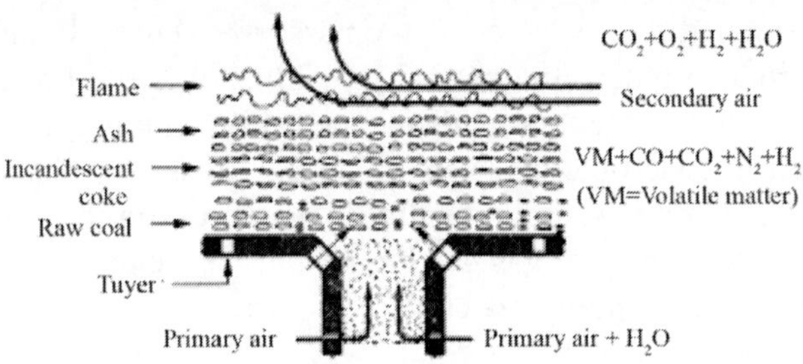

Fig. 5.2: Underfeed stoker

Grate Stokers

There are several types of biomass boilers, but the most common type is a boiler that uses grate supports for the fuel bed and to mix combustion air from beneath the grate (Figure 5.3). Biomass fuel is added at one end of the grate and is burned in a fuel bed which moves progressively down the grate, either via gravity or with mechanical assistance, to an ash removal system at the other end. A grate stoker is proven and reliable, and can tolerate wide variations in fuel quality (e.g., variation in moisture content and particle size). Moreover, they are easy to control and efficient. In more sophisticated designs, the overall combustion process is separated into three main activities

- Initial fuel drying
- Ignition and combustion of volatile constituents
- Burning out of the char

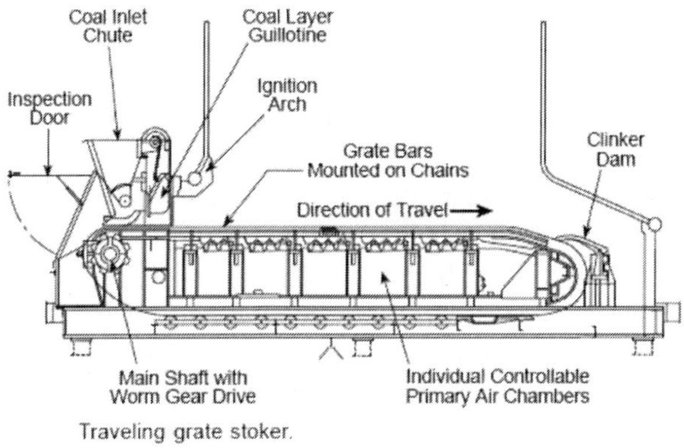

Fig. 5.3: Grate stokers

Fluidized Bed Boilers

In a fluidized bed combustion system (Figure 5.4), air is blown from below through an inert mineral bed, such as sand or limestone. An adequate volume and high enough pressure of air is pumped through the bed to entrain the small particles of the bed material so that they behave like a fluid. In fluidized bed power plants, the combustion chamber is designed so that the air velocity decreases above a certain height to allow particles to be entrainable. As a result, the bulk of the entrained bed material is retained at the bottom of the chamber. As soon as the bed becomes hot, combustible material will burn, generating heat as in a conventional furnace. Biomass is normally only 5% of the combustible material within the bed. Fluidized bed combustion is primarily driven by the reduction of SO_2 and NO_x emissions from coal combustion.

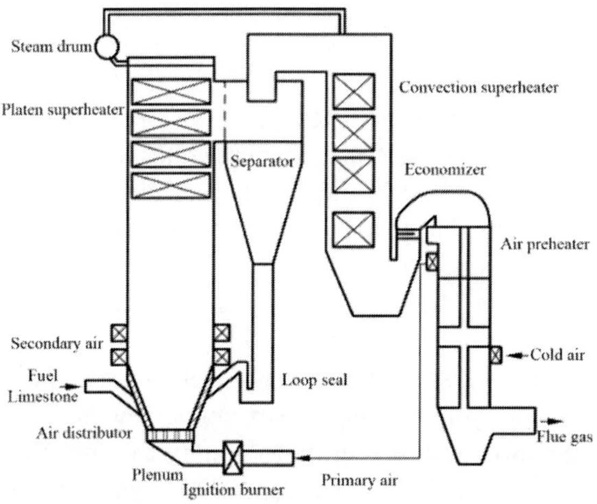

Fig. 5.4: Fluidized bed boilers

5.4 Biomass Combustion Power Plant

Most biomass power plants use direct-fired combustion systems. In a direct combustion system as shown in Figure 5.5, biomass is burned in a combustor or furnace to generate hot gas, which is fed into a boiler to generate steam, which is expanded through a steam turbine or steam engine to produce mechanical or electrical energy. It is possible to use the steam extracted from biomass power plants for manufacturing processes or for heating buildings in some biomass industries. CHP systems greatly increase overall energy efficiency from around 20% to about 80%, compared to biomass electricity-only systems. Depending on the season, the efficiency of a CHP system will differ. There are several key components to a biomass electric generation system. For a steam cycle, this includes some combination of the following items:

- Fuel storage and handling equipment
- Combustor / furnace
- Boiler
- Pumps
- Fans
- Steam turbine
- Generator
- Condenser
- Cooling tower
- Exhaust / emissions controls
- System controls (automated).

Direct combustion systems feed a biomass feedstock into a combustor or furnace, where the biomass is burned with excess air to heat water in a boiler to create steam. Instead of direct combustion, some developing technologies gasify the biomass to produce a combustible gas, and others produce pyrolysis oils that can be used to replace liquid fuels. Boiler fuel can include wood chips, pellets, sawdust, or bio-oil. Steam from the boiler is then expanded through a steam turbine, which spins to run a generator and produce electricity. In general, all biomass systems require fuel storage space and some type of fuel handling equipment and controls. A system using wood chips, sawdust, or pellets typically use a bunker or silo for short-term storage and an outside fuel yard for larger storage. Most wood chips produced from green lumber will have a moisture content of 40% to 55%, wet basis. This water will reduce the recoverable energy content of the material, and reduce the efficiency of the boiler, as the water must be evaporated in the first stages of combustion. The biggest problems with biomass-fired plants are in handling and pre-processing the fuel. This is the case with both small grate-fired plants and large suspension-fired plants. Drying the biomass before combusting or gasifying it improves the overall process efficiency, but may not be economically viable in many cases.

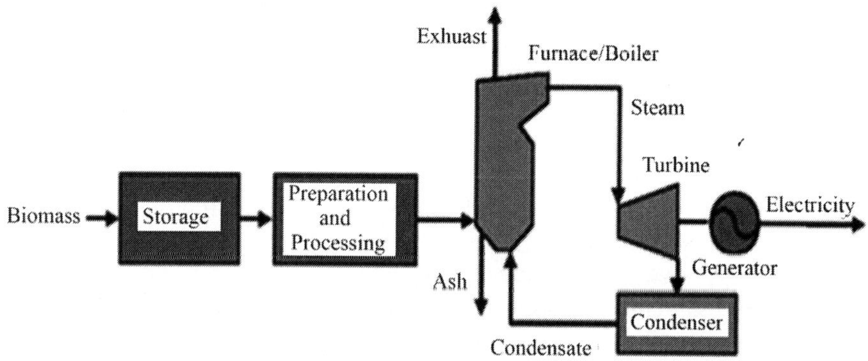

Fig. 5.5: Direct combustion steam turbine system

5.5 Benefits and Drawbacks of Biomass

Benefits

- Provides a relatively cheap fuel source
- Reduces reliance on fossil fuels
- Provides self-sufficiency, even during blackouts (except for pellet stoves that require electricity to run the auger motors and fans)
- These projects can be designed to match the electric loads as biomass can be stored and used according to demand.
- Equipment for these projects is similar to that for coal-based thermal power projects and hence, no new technological developments are required.
- Due to their proximity to the rural areas, these projects are likely to improve quality of electricity supply there.
- A variety of biomass materials can be used in the same plant, providing flexibility of operations.

Drawbacks

- Fuel quality and consistency can vary significantly
- Produces more indoor and outdoor air pollution than conventional heating fuels, even with the use of advanced combustion technology
- Creates the risk of fire and carbon monoxide poisoning
- Can require cutting wood, as well as the skill to lay fires
- Requires regular tending and maintenance
- Fuel storage requires a lot of space

5.6 Scope of Biomass Based Power Generation in India

Agricultural, agro-industrial, and forestry operations in India generate a large amount of biomass material. Agro-industrial residue alone generates over 500 million tonnes per year, according to some estimates. About 175 million tonnes of oil have the same heat content as this quantity. Some of these materials are used in the rural economy as fodder and fuel. In spite of this, studies have identified at least 150–200 million tonnes of biomass material that is not being used productively, and can be made available at a cost-effective alternative use. Straws and husks make up the majority of these materials. Electricity can be generated from this amount of biomass in the range of 15 000–25 000 MW. Further, biomass grown on wastelands, roadside plantations and rail trackside plantations can also be converted into electricity. There has been an estimate that biomass can generate in excess of 70 000 MW of electricity. Biomass could therefore generate electricity at a rate of 100 000 MW in total.

5.7 Technology Used for Biomass Based Power Generation

Biomass materials can be used to generate electricity in a similar manner to conventional coal-fired power plants. In order to generate electricity from biomass, direct combustion of biomass is commonly used. Steam drives the turbine, generating heat and electricity. The efficiency of small units of 1–3 MWe is only 10–15%, while the efficiency of large units of 10–20 MWe is 18–25%. Smaller gasification systems can have an efficiency of 20–30%, and medium gasification systems of 10–20 MWe can have an efficiency range of 38–43%. There is a wide range of combustion-based technologies available on the market today. Gasification is viewed as a technological upgrade over biomass combustion for biomass-based power generation

5.8 Biomass Gasification System

The World War II oil crisis prompted the use of wood gas generators in Europe and America. In China, wood gas generators were reported to be used commercially in 1950. The most common technology for solid biomass conversion is fixed bed gasification. Approximately 75% of all gasifiers for sale are downdraft systems, 20% are fluidized bed systems, 2.5% are updraft systems, and 2.5% are other systems. Biomass gasification is thermo-chemical conversion of solid biomass into a combustible gas mixture (producer gas) through a partial combustion route with air supply restricted to less than that theoretically required for full combustion. Typical composition of producer gas is as follows.

- Carbon monoxide - 18 % – 20 %
- Hydrogen - 15 % – 20 %
- Methane - 1 % – 5 %
- Carbon dioxide - 9 % – 12 %
- Nitrogen - 45 % – 55 %
- Calorific value - 1000–1200 kcal/m^3

When suitable internal combustion engines (ICs) coupled with generators for electricity generation are designed/adopted, producer gas can replace diesel as a fuel. Many heating applications in industry can be replaced with producer gas instead of conventional energy sources such as oil. Using biomass through gasification results in relatively clean and environmentally acceptable practices. The partial substitution of diesel in existing diesel generators (DGs) can result in huge monetary savings. There are many gasifiers available for use with wood or woody biomass; some can even be used with rice husk. It is also possible to gasify many other non-woody biomass materials, although gasifiers must be specially designed for these materials and the biomass may have to be compacted in most cases.

The biomass gasification process involves chemical reactions at high temperatures that transform biomass into combustible gas from oxygen, steam, or hydrogen. Combustible gases are used in industrial and agricultural production for a wide variety of purposes. These include the production of methanol and ethanol, the generation of energy, and the delivery of heat. Through biomass gasification, biomass can be utilized and provided with the benefits of clean fuel. Among the most efficient and cleanest ways to use biomass is through biomass gasification, which overcomes the drawbacks of biomass, such as its difficulty in burning and dispersing, and takes full advantage of the benefits of gas-fired power generation technology due to its smaller equipment and lower levels of pollution.

Biomass gasification consists of three components

1. Solid biomass is converted into gaseous fuel through gasification
2. During gasification, certain pollutants, such as ash, coke, and tar, are present in the gas and must be removed by a purification system to allow gas-fired power plants to operate properly.
3. An internal combustion engine or a gas turbine produces energy through gas-fired power generation. Using steam turbines and waste heat boilers can increase the efficiency of some processes.

5.8.1 Biomass Gasification Process

1. Agricultural and wood processing waste are used as raw materials for biomass gasification. Examples of materials to use include sawdust, tree branches, crop straw, rice husks, and corncobs. There is an abundance of materials and they are easily accessible.
2. It is in the gasifier that the gasification process takes place. There are a number of variables that influence the complexity of the reaction process, including the type of gasifier, the reaction settings, the properties of the raw materials, and the grinding size. In spite of this, they all follow the same procedure.
3. Raw materials undergo four stages of burning: thermal decomposition, oxidation, reduction, and drying. Gas quality and yield are most affected by the reduction layer.

5.8.2 Characteristics of Biomass Gasification Power Generation Technology

1. Gasification of biomass is a totally flexible technology for producing electricity. Its ability to select the appropriate power generation equipment based on the scale can ensure reasonable power generation efficiency at any scale, since biomass gasification power generation uses gas turbines, internal combustion engines, or even waste heat boilers and steam generation systems. Biomass gasification technology is easily adaptable to dispersed biomass consumption needs.
2. Gasification of biomass is a clean technology for generating electricity. As a renewable energy source, biomass can significantly reduce the

emissions of noxious gases like CO_2 and SO_2. NOx emissions can be successfully controlled since the gasification process typically produces little NOx (between 700 and 900 °C).

3. Biomass gasification is an economically viable energy source. Biomass gasification is generally one of the most cost-effective renewable energy technologies, and its flexibility allows it to be both economical and effective in small-scale applications. In addition, biomass gasification power generation technology has a smaller investment than other renewable energy power generation technologies. This is duc to its straightforward process of gas power generation and compact equipment. A small amount of conventional energy can be produced for the same cost as electricity.

5.9 Why Choose Gasification?

Energy from waste projects can be future-proofed by gasification, depending on your waste management and energy needs.

By implementing the system in a flexible, modular manner, plants can be developed more quickly and at a lower cost. As a result of the variety of waste types that can be treated, as well as the variety of products that can be manufactured, it provides a great deal of versatility. Consequently, it is able to manage a wide variety of waste types and generate a number of high-value revenue streams.

In comparison with incineration or landfilling, gasification is a more environmentally friendly method of waste disposal. In the upcoming years, there may be restrictions on new projects and stricter emission standards for incineration, which is crucial to take into consideration.

5.9.1 What are the Four Main Types of Gasifiers

The process of gasifying involves turning organic waste or fossil fuels into gases. These gases include carbon dioxide (CO_2), hydrogen (H_2), and carbon monoxide (CO). High temperatures are used to oxidize and reduce organic materials during the procedure. The finished item is a fuel or gas mixture known as producer gas or syngas. The product burns more effectively than virgin fuels because it is flammable at greater temperatures. Electricity is produced using the method in many different sectors.

5.10 Types of Gasifiers

In a gasifier, the fuel interacts with air or oxygen and steam. Therefore, gasifiers are classified according to the way in which air or oxygen is introduced. On a larger scale, there are four types of gasifiers as follows

5.10.1 Updraft Gasifier or Counter Flow Gasifier

The countercurrent or updraft gasifier is the earliest and most basic type of gasifier. The top is where the gas exits and the bottom is where the gas enters. The reduction reaction occurs further up in the gasifier after the combustion reaction, which happens close to the grate at the bottom. Due to forced convection and radiant heat transfer from the bottom zone, the feedstock is heated and pyrolyzed in the gasifier's top section. The gas stream carries the tars and volatiles created during this process. The ash is taken out of the gasifier's bottom (see Fig. 5.6).

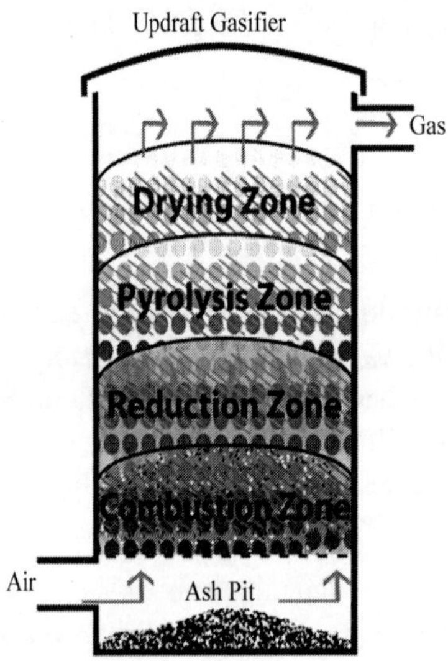

Fig. 5.6: Updraft gasifier

Advantages

a) The main advantage is its simplicity.

b) Internal heat exchange and high carbon release of gas at low temperatures.

c) High efficiency of gasification equipment.

Disadvantages

a) If there are channels in the equipment, an explosion may occur. This may be the cause of oxygen breakthrough. This is extremely dangerous and is one of the main disadvantages of this type of gasifier.

b) Tar may be burned if the gas is used in applications where it is directly heated.

c) Disposal of condensate containing tar may also be a minor problem.

5.10.2 Downflow or Parallel-flow Gasifiers

By designing down-flow or downdraft gasifiers, where primary gasification air is supplied into or above the oxidation zone in the gasifier, solutions to the issue of tar entrainment in the gas stream have been identified. In order for the fuel and gas to travel in the same direction, the generator gas is evacuated from the unit's bottom(see Fig. 5.7). In order to become the permanent gases hydrogen, carbon dioxide, carbon monoxide, and methane, the acid and tar distillation products from the fuel must pass through the flaming charcoal bed during the descent. The amount of tar that has been degraded varies depending on the hot zone's temperature and how long the tar fumes have been present.

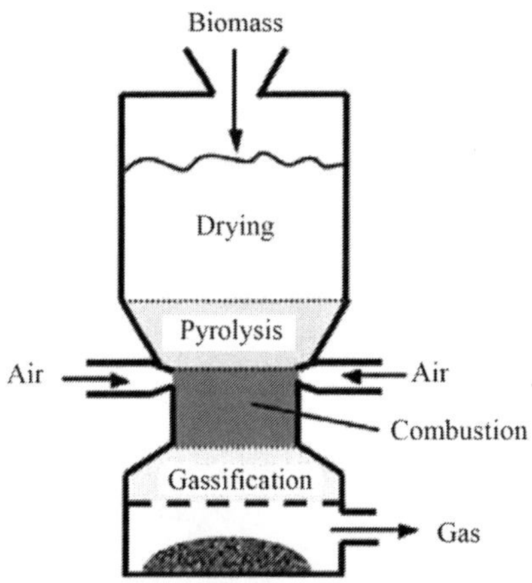

Fig. 5.7: Downdraft gasifier

Advantages

a) The product obtained is a tar-free gas that can be used in engine applications.

b) These gasifiers are less affected by any environmental obstacles due to the low content of organic components in the condensate.

Disadvantages

a) Unable to use most unprocessed fuels.

b) They encounter problems due to the use of fuels with high ash content.

c) Lack of internal heat exchange compared to updraft gasifiers.

d) Lower efficiency due to the low calorific value of the gas.

e) Gasifiers become inefficient at high uniform temperatures and may become impractical for power ranges above 350 kW.

5.10.3 Crossflow Gasifier

Cross-flow gasifiers are suitable for use with charcoal. Charcoal gasification leads to very high temperatures in the oxidation zone (1500 °C or higher), which can lead to material problems. In a cross-vented gasifier, the fuel (charcoal) itself provides insulation against these high temperatures (see Fig. 5.8).

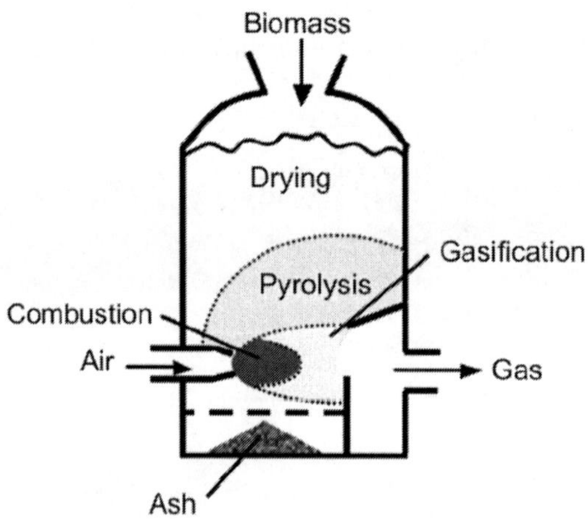

Fig. 5.8: Crossdraft gasifier

Advantages

a) These units are economically viable in the power range up to 10 kW.

b) It can be operated on a very small scale.

c) These types of gasifiers can fit into small engines.

Disadvantages

a) It requires a concomitant supply of good quality charcoal.

b) These gasifiers have a minimal tar conversion capacity.

5.10.4 Fluidized Bed Gasifier

The operation of both top and bottom suction gasifiers is influenced by the morphology, physical and chemical properties of the fuel. Common problems are the absence of fuel flow, slagging and extreme pressure drops on the gasifier. One design approach designed to eliminate these difficulties is the fluidized bed gasifier. Air is blown through the bed of solid particles at a rate sufficient to keep them in suspension (see Fig. 5.9). The bed is initially heated from the outside and once a sufficiently high temperature is reached the feedstock is introduced. Fuel pellets are introduced to the bottom of the reactor and are quickly mixed with the bed material, which is heated almost instantaneously to bed temperature. As a result of this treatment, the fuel is pyrolyzed very rapidly, resulting in the mixing of components with relatively large amounts of gaseous material. Further gasification and tar conversion reactions take place in the gas phase. Most systems are equipped with an internal cyclone to minimize coke ejection.

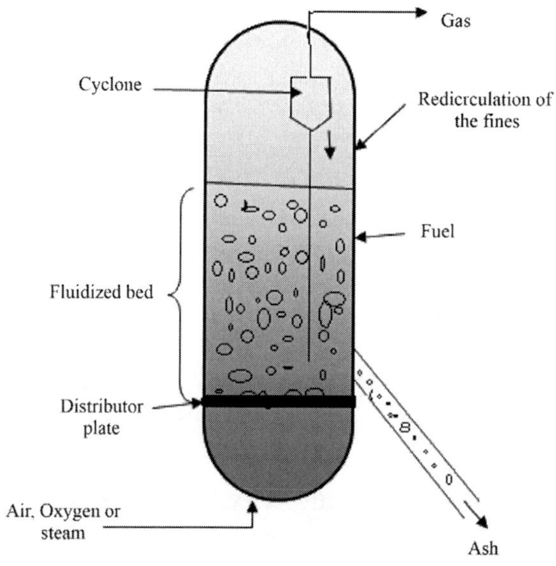

Fig. 5.9: Fluidized Bed Gasifier

Advantages

a) Feedstock flexibility as the temperature is easily controlled and the temperature can be kept at or below the melting point of the ash (rice husk).

b) Ability to handle fluffy and fine-grained materials (sawdust, etc.) without pretreatment. Some biomass fuels may have feeding problems, bed instability and fly ash sintering in the gas channel.

Disadvantages

a) High tar content of the gas (up to 500 mg/m^3 gas), incomplete carbon combustion and poor response to load changes.

b) There are no very small fluidized bed gasifiers, and the application range must be tentatively set at 500 kW (shaft power) or more.

c) Fluidized bed gasifiers are currently available on a semi-commercial basis from a number of manufacturers in Europe and the USA.

5.11 Factor Affecting Gasification Process

a) Fuel type

b) Fuel size

c) Solid waste

d) Form

e) Moisture and ash content

f) Oxygen flow in the gasifier

5.12 Advantages and Disadvantages of Gasification

Advantages

(1) During pyrolysis gasification, the organic components in the waste can be converted into different forms of usable energy such as combustible gas and tar, and its economy is better.

(2) The air coefficient is lower during pyrolysis gasification, which greatly reduces the amount of smoke emission, improves energy utilization, reduces ammonia oxide emission, and reduces the investment and operation cost of flue gas treatment equipment

(3) Under the reducing atmosphere, metals are not oxidized, which is convenient for recycling, and Cu, Fe and other metals are not easily generated to promote the formation of dioxins catalyst

(4) The content of heavy metals, dioxins and other pollutants in the flue gas produced by pyrolysis gasification method is less, and the secondary pollution is smaller, so the pollution control problem is simplified and safer for the environment.

Disadvantages

(1) It is a misunderstood technology, and a serious problem with gasification is that many people in the market do not understand it.

(2) Due to some of the failures of gasification projects – especially in the UK – some people believe that the technology is not applicable on a commercial scale or for certain types of waste materials. Others believe that the technology is not viable.

5.13 Capacities of Gasifier Based System

Biomass gasifier-based systems are being made in capacities ranging from a few kilowatts to a megawatt of electricity equivalent. For heating applications, the current upper limit on the unit size is equivalent to 200–300 kg/h of oil consumption. A representative of biomass gasifier based power generation system is shown in Figure 5.10.

Costs

The typical costs of biomass gasifier-based electricity generation systems range from Rs 4 crores/MWe to Rs 4.5 crores/MWe. The cost of power generation depends on cost of biomass, plant load factor, etc., and is estimated to be between Rs 2.50/kWh and Rs 3.50/kWh. For thermal applications, the capital costs are estimated to be about Rs 0.5–0.7 crores for each 1 million kcal capacity.

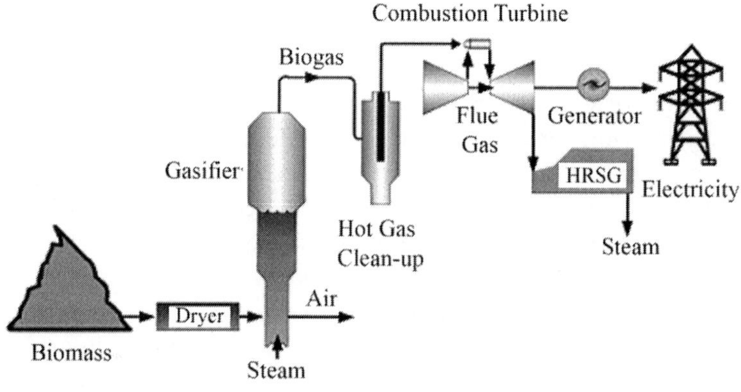

Fig. 5.10: Gasification based power plant

5.14 Scales of Biomass Power Generation

A biomass gasification power generation system can be classified as small, medium, or large, depending on its size.

1. Small biomass gasification power generation systems serve mainly as self-sufficient generator sets for small and medium-sized businesses or as rural lighting systems. They can select the best gasification equipment for the various shapes of biomass, with a general power output of 200 kW, since they only need a small quantity of one type of biomass.
2. It is possible to use one or more different types of biomass in a medium-sized biomass gasification power generating system, but a larger quantity of biomass must be used. Large and medium-sized industries typically use it for self-supplied power or as a small-scale feed-in station. In most cases, fluidized bed gasification is used, and pretreatments such as crushing and drying are required. There are several types of biomass gasification technology, but the most common type is the medium-sized biomass gasification power production system, which can generate electricity between 500 and 3000 kW and is widely used.
3. Large-scale biomass gasification power generation systems function primarily as feed-in power stations that can use a wider range of biomass and require significant amounts of biomass. Additionally, future biomass utilization requires the support of specialized biomass supply centers and pretreatment centers. Even though biomass gasification is still small when compared to conventional energy, big biomass gasification power generating systems are often over 5000 kW, and as the development of biomass energy matures, it will become a major source of power to replace conventional energy.

5.15 Biomass Pyrolysis

Pyrolysis occurs when biomass is heated without oxygen and decomposes. A chemical reaction that occurs naturally within the first two seconds of combustion and gasification precedes this reaction. Methane, hydrogen, carbon monoxide, and carbon dioxide are some of the gases produced by biomass pyrolysis. As pyrolysis takes place at either low temperatures below 450 0C or high temperatures above 800 °C, depending on the thermal environment and end temperature, it will produce biochar between the two. Generally, at low temperatures below 450 °C, biochar will be produced slowly. At high temperatures above 800 °C, gas will be produced primarily, where there will be rapid heating. The main product is bio-oil when temperatures are moderate and the heating rate is relatively high.

By increasing the energy density of biomass resources and reducing transportation and handling costs, pyrolysis can be performed at relatively small scales and in remote areas. Pyrolysis is a process that absorbs heat, so it requires sufficient heat transfer surfaces to handle the temperature requirements. A variety of chemicals, heat, and electricity can be produced using pyrolysis, which is a flexible and attractive way to convert solid biomass into liquids.

5.15.1 Three Types of Biomass Pyrolysis

(1) **Slow pyrolysis:** Biomass is pyrolyzed at very low temperatures for long periods of time (hours or days) to obtain a maximum coke yield of 35% (mass fraction) for charcoal burning, also called charring of biomass.

(2) **Conventional pyrolysis:** The biomass feedstock is put into a conventional pyrolysis unit and pyrolyzed at less than 500°C, lower heating rate (10~100 K/min) and 0.5~5s residence time of pyrolysis products, which can make the same proportion of gas, liquid and solid products (20%~25% biochar and 10%~20% bio-oil by weight of feedstock can be obtained).

(3) **Rapid pyrolysis:** Fast pyrolysis is to put finely ground biomass raw materials into the fast pyrolysis device, biomass at atmospheric pressure, ultra-high heating rate (1000~10000 K/s), ultra-short product residence time (0.5~2s), moderate pyrolysis temperature (500~650 □) instantaneous gasification, and then rapid condensation into liquid, can obtain the maximum liquid yield, the product of bio-oil can generally reach The heat required for the rapid pyrolysis process is partly generated by the pyrolysis gas as the heat source.

5.16 Economic and Environmental Advantages of Pyrolysis

1. The potential environmental benefits of using renewable resources in carbon-neutral ways
2. Utilization of waste materials, such as wood processing waste (bark, sawdust, deforestation, etc.), agricultural residues (straw, fertilizer, etc.) – economic potential
3. Self-sustainable energy – economic potential
4. Conversion of low energy in biomass into high energy density liquid fuels – environmental and economic potential
5. Potential for production of chemicals from biological resources – environmental and economic potential

5.17 Utilization of Pyrolysis Products

The main products of biomass pyrolysis are bio-oil, non-condensable gas, and charcoal. Among its uses, bio-oil is primarily used in combustion for heating, electricity production, fuel oil, and chemical production. As bio-oil has a lower calorific value than fossil fuel oil and has a high moisture content, its application is more difficult. However, liquid products have the advantages of being easy to handle, transport, and store, as well as allowing existing fuel oil installations to be used. High viscosity and deterioration susceptibility of biomass oil pose the main problem in the application of bio-oil for combustion. It is possible to separate the production of a liquid product from the production of electricity when producing a liquid product, so peaking power stations can be equipped with pyrolysis units or liquid products can be sent directly to electricity-generating central power stations. Bio-oil contains many chemicals and products, and hundreds of substances have been identified in the liquid product of pyrolysis, so there is a great deal of interest in recovering compounds or congeners from it, and the potential value of some specific compounds, even if recovered in small amounts, is much higher than the value of recovered oil. In conclusion, the application of bio-oil as a fuel in boilers, turbines and diesel engines is relatively successful experimentally, and its refinement, purification and extraction of chemical substances are not yet mature in terms of research and do not have economic feasibility.

5.18 Biomass Pyrolysis Process Considerations

A pyrolysis unit requires several components in addition to the pyrolyzer.

(1) **Production of bio-oil from the pyrolyzer.** Bio-oil is refined and fractionated so that it can be used as a transportation fuel. Specialty chemicals can also be made from them, particularly cyclic structured compounds used in adhesives. Due to the presence of combustible components, these gases generate heat. Additionally, biochar is produced. As a soil conditioner, biochar can improve soil quality, and it can also be used as a catalyst carrier or activated carbon. In addition to ash, a mineral material is produced in the process. Ash must usually be included.

(2) **The next unit to consider is the separation unit.** Since carbon is a solid, it is usually separated using a cyclone or baghouse. Due to the minerals in charcoal and the fact that carbon catalyzes the gasification reaction, it can be used as a catalyst for further decomposition into gas. Separating the liquid from the gas is also necessary. In order to separate condensable liquid from non-condensable gas, the liquid and gas must

be cooled. In order to improve the liquid's stability, it is then fractionated and likely further processed. In some cases, heavier components can clog the liquid fraction. When possible, non-condensable gases can be reused after they have been cleaned of any trace liquids.

(3) **The next consideration is the heat source for the equipment.** The hot flue gas is used to dry the feed material. Due to the presence of combustible gases, flue gases can be partially burned to generate heat. In addition to burning any remaining coke, there will be a secondary heat source. As another major source of heat, biomass can also be partially burned.

(4) **Another important process to consider is the mode of heat transfer.** The majority of them are indirect, using metal walls and tube and shell units. Char and biomass combustion involve direct heat transfer. Heat is provided to the pyrolyzer from the carrier (usually sand) in fluidized bed systems.

Advantages and disadvantages of combustion

Advantages

(1) Incineration is a proven technology that has proven to be successful in the marketplace. More than 500 incinerators operate in Europe, and the European Federation of Energy from Waste Plants represents 410 of the continent's 23 nations. Over the past two decades, the amount of rubbish burned in Europe has more than doubled to 58 million tons.

(2) A straightforward trash disposal technique: incineration can burn up to 90% of solid waste, making it more efficient than landfills.

(3) Reduced landfills: According to the European Environment Agency, the percentage of trash disposed of in landfills decreased from 29% to 24% between 2010 and 2016, and it has continued to decline since then. This drop can be partly attributed to the increase in incineration. As a result of government efforts to limit landfills, incineration will soon account for over half of all garbage disposed of in the United Kingdom.

(4) Energy production: Incineration can produce energy and electricity as a result of the heat produced.

Disadvantages

(1) Expensive. Incineration plants require a lot of infrastructure to operate. Additionally, incineration plants require trained personnel and dedicated

staff to operate. In addition to high operating costs, incinerator plants require regular maintenance.

(2) Pollution of the environment. The combustion process in an incinerator produces smoke. There are acidic gases, carcinogens, dioxins, particulates, heavy metals, and nitrogen oxides in the smoke. The environment is adversely affected by these gases. A number of studies have shown that dioxins produced by plants cause cancer.

(3) Potential for long-term problems: Incineration does not encourage recycling and waste reduction. This is not a calculated strategy for any society. The focus should be on reducing waste and recycling most of the waste. Simply burning most of the waste without recycling some of it will only cause further environmental damage, as it may encourage more waste generation.

(4) Ash waste can be hazardous to humans and the environment: Although the amount of ash remaining from the process is relatively small, it contains many toxins and heavy metals that require further treatment. If not properly configured, it may pose a serious hazard to the public and the environment.

5.19 Bagasse Based Cogeneration

The process of cogeneration involves using a single fuel to produce more than one form of energy. Steam and electricity cogeneration can significantly improve the overall efficiency of fuel utilization in process industries. Cogeneration (Figure 5.11) requires the simultaneous production of heat and electricity in a favourable ratio, which is well met in the sugar industry. During electricity production, a large amount of heat is rejected to a lower temperature sink. Normally, up to 70% of the heat in steam is rejected to the atmosphere in condensers in electricity generation plants. Cogeneration, however, employs this heat to meet process heating needs rather than wasting it. Sugar mills in the country have the potential to generate 3500 MW of additional electricity through cogeneration. There are three main pieces of equipment required for these projects: bagasse-fired boilers, steam turbines, and a grid-interphasing system. Indigenously manufactured equipment is used in all of these applications. Fuel utilization can thus be improved by 60% or even more in some cases. In cogeneration projects, electricity is produced at a rate ranging from several kilowatts to several megawatts, and the heat produced at a rate ranging from a few kWth to several MWth. It is estimated that the capital cost of bagasse-based cogeneration projects in sugar mills ranges between Rs 3 crores/MW and Rs 4 crores/MW. A typical sugar mill with an average crushing season of 160 days will benefit in the long run from investing in additional power generation through cogeneration.

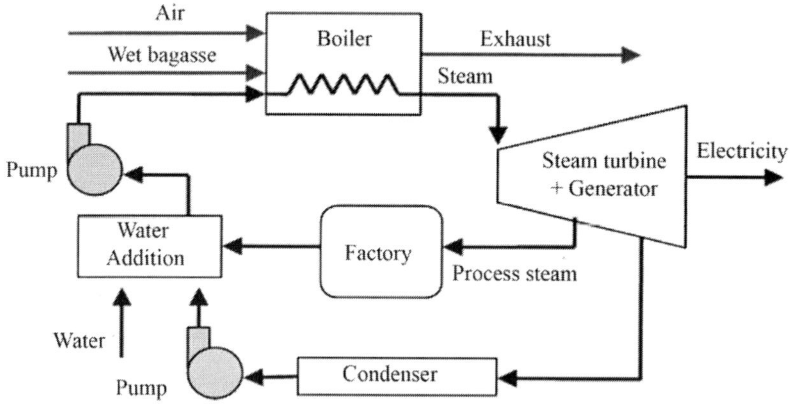

Fig. 5.11: Cogeneration using bagasse

5.20 Efficiency of Biomass-based Power Generation

Plants that generate electricity from biomass can achieve an efficiency between 20 and 40%. Direct combustion steam plants (up to 1 MWe) have low efficiency, but IGCC plants at medium scales (over 20 MWe) can reach as high as 40% conversion efficiency. Boilers can achieve an efficiency of 65–90%, while steam turbines (50–60 bar and > 550°C) can achieve 35–55%. This results in an overall conversion efficiency of 35–45% for a cogeneration system. Boiler efficiency and turbine efficiency are both factors that determine the overall efficiency range. Combining an efficient boiler with a high-efficiency turbine results in a high combined efficiency. An overall efficiency of 45 to 50% is achieved when the power cycle consists of two cycles (gas turbines and steam turbines or high-pressure and low-pressure turbine sets). In the commercial sector, sustainable biomass production is an important source of power production. Furthermore, biomass can also create additional employment opportunities. Biomass industrialization faces three critical challenges: mass-scale sustainable biomass production, pretreatment of bulk biomass for higher calorific value, and scaling up of plant capacity for commercial viability.

Multiple Choice Questions

1. The process of usage of organic material to generate energy is known as

   a) Bio mass energy
   b) Hydro thermal energy
   c) Geothermal energy
   d) Nuclear energy

2. The term biomass most often refers to __________

   a) Inorganic matter
   b) Organic matter
   c) Chemicals
   d) Ammonium compounds

3. What type of energy is biomass energy?

   a) Conventional energy
   b) Non renewable
   c) Commercial energy
   d) Sustainable energy

4. Biomass is useful to produce __________

   a) Chemicals
   b) Fibers
   c) Biochemicals
   d) Transportation fuels

5. The aerobic digestion of sewage is used to produce __________

   a) Biomass
   b) Bio fuels
   c) Synthetic fuels
   d) Metal articles

6. Which of the following technologies are used to convert biomass into useful energy forms?

   a) Bio-chemical process
   b) Galvanization
   c) Doping
   d) Photoelectric effect

7. What is unique about the gasification agent entering in a fluidized gasifier?

   a) Enters from bottom at a relatively fast rate as compared to a fixed bed gasifier
   b) Enters from bottom at a relatively slow rate as compared to a fixed bed gasifier
   c) Enters from top at a relatively fast rate as compared to a fixed bed gasifier
   d) Enters from top at a relatively slow rate as compared to a fixed bed gasifier

8. What ensures consistency of temperature in fluidized gasifier?

   a) Gasification agent

   b) Air-fluidized bed material

   c) Biomass

   d) Air non-permeable bed material

9. How is the temperature maintained in a bubbling fluidized bed?

   a) Biomass

   b) Gasification agent

   c) A constant ratio of biomass and gasification agent

   d) Manipulating the ratio of biomass and gasification agent

10. Which of the following is a product of biomass gasification?

    a) Hydrogen b) Steel

    c) Carbon (solid) d) Iron

11. What are the three main types of gasifiers?

    a) Fixed bed, hydrothermal liquefaction and carbonisation

    b) Fixed bed, fluidized gasifiers and carbonisation

    c) Carbonisation, liquefaction and entrained flow gasifiers

    d) Fixed bed, fluidized gasifiers and entrained flow gasifiers

12. How is the biomass material and gasification agent fed into an updraft gasifier?

    a) Biomass from top, gasifying agent from top

    b) Biomass from top, gasifying agent from bottom

    c) Biomass from bottom, gasifying agent from top

    d) Biomass from bottom, gasifying agent from bottom

13. The product gas removed from the low temperature zone undergoes _______ before being used as fuel in combustion for electricity generator.

    a) Liquefaction b) Condensation

    c) Evaporation d) cleaning

14. The tar content of the product gas in downdraft gasifier is _______ updraft gasifier.

a) equal to b) less than

c) greater than d) cleaner than

15. Which of the following applications is the product gas from downdraft gasifier suitable for?

a) Fuel for combustion engine

b) Fuel for burning wood

c) Fuel for internal combustion engine

d) Fuel for household purposes

16. How does air enter and exit in a cross-draft gasifier?

a) Air enters from one of the sides and exits from the top

b) Air enters from one side and exits from the other

c) Air enters from one of the sides and exits from the bottom

d) Air enters from the bottom and exits from the top

Answer

1	a	2	b	3	d	4	d	5	b
6	a	7	a	8	b	9	d	10	a
11	d	12	b	13	d	14	b	15	c
16	b								

6

Biomass Pelletization and Baling

6.1 Introduction

Biomass pellets are one of the most common types of biomass fuel. It is generally composed of wood waste, agricultural biomass, grasses, and forest residues, and is one of the most popular types of biomass fuel available today. Pelletization of biomass results in significant savings in terms of transportation, and storage, as well as the ease of handling biomass as a whole. In terms of flowability, dense cube pellets are similar to cereal grains in terms of their characteristics of being dense and cube-shaped. Due to their regular geometry and small size, biomass pellets are easily fed through an automatic feed system with fine calibration thanks to their regular shape. Because pellets are so dense, they can be stored in a compact way and transported over a long distance in a rational and efficient manner, because of their density. Furthermore, pellets can also be produced with a low moisture content, even though they are extremely dense. This allows them to burn with a high degree of combustion efficiency due to their low moisture content.

The pelletization of biomass has long been known as an established process for making solid energy carriers from biomass. These carriers have a high density and are composed primarily of biomass. There are various types and grades of pellets that can be used as fuel in homes, electric power stations, and other applications as solid fuel. Due to a number of pellet-making equipment that is available in a variety of sizes and scales, it is possible for pellets to be made on both a domestic and an industrial scale, depending on the equipment. Approximately 6-25 mm in diameter and 3-50 mm in length, the pellets are cylindrical in shape with a diameter of 6 to 25 mm.

6.2 Pelleting Process

There is a method of making pellets that involves placing ground biomass under high pressure and forcing it through an opening in a die that is shaped like a circle. The biomass fuses together when exposed to the appropriate conditions, forming a solid mass when exposed to the appropriate conditions. It is natural for biomass (mostly wood) to be able to form high-quality fuel pellets by itself. Nevertheless, other types of biomass may need additives to act as binders in order to keep the pellets together so they can be molded into pellets. Extrusion is the process that is used to carry out this process.

However, the creation of pellets is only a small part of how oil and gas pellets are produced as a whole, which is only one of the many steps. This process entails grinding the feedstock, controlling the moisture in the feedstock, extrusion, cooling, and packaging the end product. A final product of acceptable quality must be produced carefully at each step in order to achieve the desired outcome.

A biomass pellet mill of a standard size typically requires its biomass to be ground to a particle size of no more than 3 mm in order to be able to produce pellets. It is possible to perform this task with a wide variety of equipment. A hammer mill or similar device is used to reduce smaller particles from large and dense biomass (such as wood), which is first processed through a chipper, and then through a hammer mill. Biomass that is smaller and softer (such as straw) can be fed directly into the hammer mill without having to be chipped first.

Pellets are actually created through the process of extrusion. In the compression process, biomass is compressed against a heated metal plate called a "die" that has been drilled with several small holes, which allow the biomass to be squeezed through under high temperatures and pressures using a roller. In the right conditions, biomass particles will fuse into a solid mass, resulting in pellets being formed when the conditions are just right. It is usually a blade that is used to cut the pellet as it exits the die to a predetermined length as it exits the die. There are some cases where biomass that fuses together better than others has a tendency to do so in some cases. It is thought that the lignin in sawdust acts as a glue in pellets, which makes it a very suitable feedstock for pellet production since it acts as a glue to hold pellets together. Due to the fact that grass pellets do not fuse as well as grass pellets, they are less dense and easier to break than grass pellets. This problem can be minimized or eliminated if input material properties are combined with the operation of pelleting equipment. Additionally, in order to help the biomass stick together, it is possible to mix a fraction of sawdust with the biomass in order to achieve similar results. Biomass can be improved in some cases by distillers dry grains, which are a product of the corn ethanol industry.

As pellets leave the die, they are quite hot (150°C) and fairly pliable (being soft just after they leave their die). For this reason, they are ready for use only after they have been cooled and dried. It is usually accomplished by blowing air through the pellets while they are sitting in a metal bin. Pellets should not exceed 8 percent moisture content at the end of the process. Schematic of pellets manufacturing process is illustrated in Fig 6.1

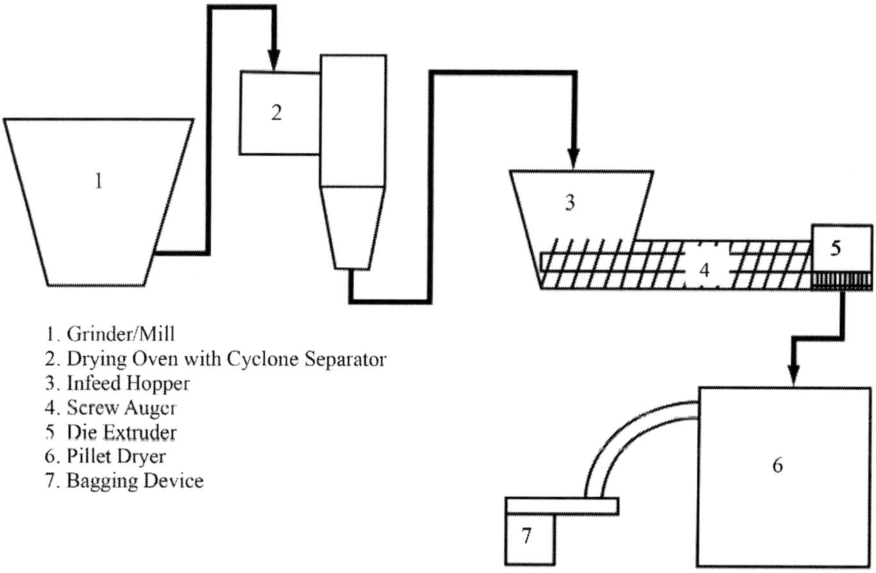

Fig. 6.1: Pellets making process

An overhead hopper and a conveyor belt are used to pack the pellets into bags. Pellets are stored in elevated storage bins or in silos at ground level. Pellets should be packaged in a manner that prevents moisture and pollutants from entering. The availability of commercial pellet mills and pelletizing equipment is widespread throughout the world.

As lignin, which is naturally found in wood and is recognized as one of the most important substances in the production of wood pellets, particles rise to the surface of the pellets during production and this creates a smooth and shiny surface on the pellets. Furthermore, lignin serves as a natural glue that holds pellets together. It is important to note that pellets are composed of 100% wood, and there is no other binding agent in them other than the natural lignin of the wood.

6.3 Calorific Value of Different Pellets

The type of biomass material will determine the pellets calorific value, and we made a summary of the calorific value of pellet fuel based on the type of biomass material: the calorific value of agricultural raw material pellet fuel typically ranges between 12.55-17.60 MJ/kg, whereas the calorific value of forestry raw material pellet fuel generally ranges between 17.60 and 21.00 MJ/kg, and after carbonization, forestry pellet fuels are capable of producing a calorific value of no less than 30 MJ/kg as presented in Table 6.1.

Table 6.1: Calorific value of pellets of different biomass

Different Pellets	Calorific Value
Weed Pellets	16.20 MJ/kg
Rice Husk	16.01 MJ/kg
Rice Straw	17.63 MJ/kg
Corn Stalk	4241 MJ/kg
Corn Cob	17.74 MJ/kg
Wheat Straw	18.52 MJ/kg
Peanut Shells	21.41 MJ/kg
Chinese Fir	19.19 MJ/kg
Pine Wood	19.10 MJ/kg
Poplar Pellets	17.93 MJ/kg
Cow Dung	11.62 MJ/kg
Firewood	16.75 MJ/kg
Tree Leaf	14.62 MJ/kg
Bean Pole	16.15 MJ/kg
Sorghum Stalk	15.10 MJ/kg
Sophora Stick	18.10 MJ/kg
Wood Chips	18.41 MJ/kg
Furfural Residue	13.15 MJ/kg

6.4 Factors Affecting Pellet Quality

6.4.1 Type of Raw Materials

Different types of raw materials have different pelletizing processes that vary greatly depending on the type of raw material. In spite of the large quantities of agricultural and forestry waste there are, some of them are easier to crush and shape into pellets after they have been crushed. Raw materials affect the quality of pellets forming, such as density, strength, calorific value, etc., and have a certain impact on the output and power consumption of biomass pellet mills.

6.4.2 Raw Material Size

The particle size of the raw material is also one of the most important factors affecting the pellet forming process. It is generally accepted that the finer the size of the raw material, the higher the strength of the pellet will be, but it is not suitable to adjust humidity and is easy to agglomerate. It is also important to point out that if the raw materials are crushed too finely, the crushing process will consume a lot more energy than it needs to.

There is a possibility that the die and the pressure roller will wear and tear more quickly if the particle size of the raw material is too coarse. As a result of their hardness to form, pellets are difficult to form, especially if the die has a small diameter. Therefore, as a result of this, the softening paste effect is poor, leading to a high material consumption, a low yield, and a high powder content in the final pellets, as a result.

6.4.3 Moisture Content of Raw Materials

Friction between raw materials will limit sliding and extension when the moisture content of raw materials is too low. Thus, pellets formed will not be firm or even hard to shape when raw materials are not close enough; when moisture content is too high, even though the flow of raw material particles is good, they can be fully extended, and they can engage in mutual engagement. When raw materials are extruded, excess water will be distributed between them. This will make it difficult for the raw materials to fit closely, causing the form to be unstable, or even difficult to shape.

There are different humidity requirements for different raw materials when producing pellets, depending on temperature and other factors. Most raw materials require 8% to 15% humidity, but there are some special materials that require higher humidity, such as teak wood, which requires even up to 50% humidity. As this affects the quality of the final pellets, it's better to pre-test them first. Raw materials with excessive humidity must be dried before pelletizing to reduce moisture.

6.5 Government Initiatives

Efforts are being made to meet the demand for 5% co-firing in Thermal Power Station across the country. With the Government policy in force and multiple initiatives taken to strengthen supply chain infrastructure, pellet/briquette manufacturing capacity is expected to rise in near future.

About 23 Thermal Power Plants have co-fired biomass pellets so far. Approximately 66000 Metric Tons (MT) of biomass has been co-fired till date.

As per present initiative, Biomass Pellets are targeted to co-fire with coal. The blending of biomass with coal will reduce coal dependence in thermal power plants.

Biomass co-firing in Thermal Power Station (TPS) has several benefits as under:

a) Reduction in amount of coal used for power generation resulting in corresponding savings in CO2 emissions. This would also reduce the sector's dependence on coal.

b) Production of electricity from a resource (biomass) which was earlier being wasted.

c) Income generation for farmers and job creation by pellet manufacturers.

d) Preservation of soil culture which gets destroyed in farm fires.

e) Reduction in air pollution due to reduction in stubble burning.

6.6 Balers for Agricultural Crop Residues Management

An agricultural baler is a machine that compresses and scrapes cut and scraped crops into a compact or suitable size that can be stored. A baler attached to a tractor can be used to convert agricultural waste, such as straw and hay, into bales that can be used as animal feed and haylage.

The use of balers makes it easier to transport and store crop residue. This can result in an 80 percent reduction in trash production, allowing you to save time, money, and space. In addition, it reduces the risk of fires.

6.7 Baler Machine

As the name suggests, a crop residue baler machine is a machine that collects crop residue and bundles it for disposal. It can be used to collect crop residue from a variety of crops, such as paddy, jowar, wheat, cotton, millet, and many others. It is a PTO-powered machine.

6.8 Functioning of a baler

- Bales are formed by first picking up crop residue with pick-up units and then rolling it with rollers.
- Once the bale is ready, the tractor driver stops and ties it with a rope. The bales retain their shape as a result of tying.
- A bale indicator mounted on the baler informs the driver of what is taking place inside the machine.
- A bale ejection switch is located in the tractor's control box after the bale has been formed.
- As soon as the driver presses the eject switch on the baler, the door opens and the bale emerges from the baler.

6.9 Types of Baler Machines in Agriculture

It is generally accepted that balers are classified according to the kind of material to be baled, the loading technique, the purity required, the dimensions of the transport and milling process, and the throughput requirements.

6.9.1 Vertical Balers

This type of baler compresses materials by applying vertical, downward pressure to the material. Occasionally, there are downstroke balers, which are also known as vertical balers. In most cases, this equipment is used for recycling cardboard, plastics, metals, and foam, which are the most common materials.

6.9.2 Horizontal Balers

Horizontal balers compress materials horizontally by compressing them horizontally. Material collected by a horizontal baler is placed in a hopper at the top of the machine. This is in contrast to a vertical baler, where materials are directly placed into the chamber of the machine.

6.9.3 Two Ram Balers

In a two-ram baler, also known as a two-cylinder baler, one ram is used to compress the material in order to compact it. A second ram serves as a control mechanism. It is possible for them to be oriented vertically or horizontally. Rubber and other materials with high rebound properties can be compacted with these tools.

7

Biodiesel Production

7.1 Introduction

A biodiesel fuel is an environmentally friendly, renewable, biodegradable fuel that is produced at domestic facilities from vegetable oils, animal fats, or recycled grease from restaurants. In accordance with the Renewable Fuel Standard, biodiesel meets the requirements for biomass-based diesel as well as the requirements for advanced biofuels as a whole. Biodiesel is distinct from renewable diesel, also known as "green diesel." In its pure, unblended form, biodiesel is also known as B100 or neat biodiesel. Biodiesel is used to fuel compression-ignition engines in the same way as petroleum diesel. Alternatively, biodiesel is a liquid biofuel that can be used in diesel engines alone or blended with diesel oil. The fuel is obtained through the chemical conversion of vegetable oils or animal fats combined with alcohol.

Biodiesel is defined as a mixture of long-chain monoalkylic esters derived from fatty acids obtained from renewable sources, and intended for use in diesel engines by ASTM International (formerly known as the American Society for Testing and Materials).

The percentage of biodiesel in blends with diesel fuel is indicated in brackets as 'Bx', where 'x' indicates the percentage of biodiesel. As an example, 'B5' denotes a biodiesel blend containing 5% biodiesel and 95% diesel fuel; in consequence, 'B100' denotes pure biodiesel.

7.2 Raw Materials for Biodiesel Production

Biodiesel is now predominantly produced from biological sources, such as vegetable seed oils, soybean oils, and some recovered animal fats. In comparison to conventional diesel fuel, biodiesel is a clean, renewable resource that is biodegradable, nontoxic, and nonpolluting.

Globally, soybean, palm, sunflower, and rapeseed oils are the most common oils used for biodiesel production, but peanut, linseed, safflower, and used vegetable oils are also used.

In order to reduce costs in the biodiesel production and trade process, research has been conducted for several years on the use of non-edible vegetable oils for the production and sale of biodiesel.

In addition to being cheaper than the use of edible oils in the production of biodiesel, the use of non-edible oils for the production of biodiesel has the advantage of not requiring the use of food. The production of biodiesel has been tried in a number of countries through the use of castor oil, tung oil, cotton, jojoba, and jatropha oil, as well as other non-edible oils. Furthermore, the use of animal fats is also an option in countries with a high level of livestock resources. However, they should be pretreated because of the solid nature of their fats. Other alternatives include high acidity grease from animals including cattle, pork, poultry, and fish.

Despite their high oil yield, microalgae appear to be a promising alternative to producing biodiesel in the future. Nevertheless, it is pertinent to note that there are a few species that are more suitable for producing biofuels than others. Despite the differences in the properties of oils and fats as raw materials, biodiesel must meet international standards.

7.3 Biodiesel Production Process

The most common feedstocks used in the production of biodiesel are recycled vegetable oil, virgin vegetable oil, and tallow. It should be noted that recycled oil is processed to remove impurities such as dirt, charred food, and water from the process of cooking, storing, and handling the oil. It should be noted that virgin oil is refined, but not to a level at which it would be suitable for human consumption. There are many ways of refining rice, but during the refinement process, phospholipids, as well as other plant matter, are typically removed by degumming. During the process of base-catalyzed transesterification, the triglycerides are saponified (hydrolyzed), which is a process that produces soap instead of biodiesel when there is water present.

To determine the amount of free fatty acids present in the vegetable oil sample, we titrate a sample of the cleaned feedstock against a standardized base solution. This will enable us to determine the amount of free fatty acids present. The production of biodiesel is either done by neutralizing the acids or by esterifying them to produce biodiesel.

7.4 Transesterification for Biodiesel Conversion

Biodiesel is made by using a chemical reaction in which triglycerides (fats) derived from oils (feedstocks) are transformed into the fuel biodiesel. There is no doubt that biodiesel produced by transesterification has a much lower viscosity than petroleum diesel. This makes it a suitable replacement for petroleum diesel in diesel engines since it has a lower viscosity.

In industries around the world, biodiesel is produced at industrial levels by alkoolysis (transesterification) of triglycerides, which are the main components of vegetable oils and animal fats. Due to the fact that methanol is usually the least expensive form of alcohol, methyl esters are the most common type of esters. This is because they have the simplest structures. As a result of the low cost of ethanol in Brazil, ethanol esters are the most commonly produced esters due to the low cost of ethanol. At either a low or high temperature, the reaction can be catalyzed by bases, acids, and enzymes, and it can occur at either a low or high temperature. During the process of methyl esters of fatty acids, glycerol is produced as a by-product of the reaction.

The reaction is accelerated by the use of a catalyst. Depending on the catalyst used, transesterification can be either basic (most frequently used at all production scales), acidic (rarely used in industrial production, usually used as a first stage when the raw materials are highly acidic), or enzymatic (less commonly used; the enzymes are usually lipases), with the former being the most common method. In transesterification, R', R" and R"' are chains comprising carbon atoms and hydrogen atoms (chain fatty acids) as shown in Fig 7.1.

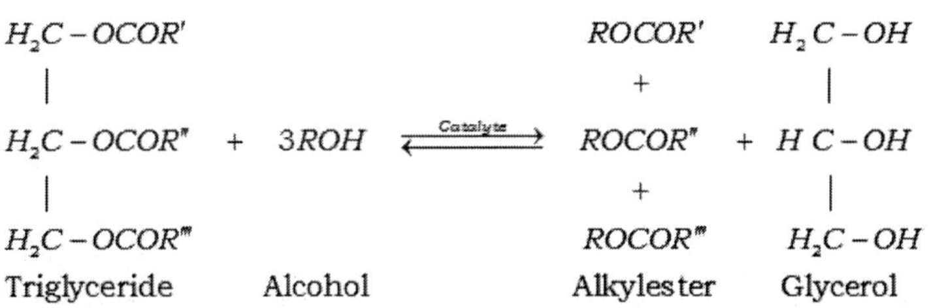

Fig. 7.1: Transesterification reactions of triglycerides

As a result of mixing chemicals with vegetable oil, two essentially immiscible phases are formed. One phase is apolar in nature, containing triglycerides and esters, and the other phase is polar in nature, consisting of glycerol and alcohol. A saponification reaction occurs when an alkali catalyst is used. During this process, two immiscible phases form, resulting in the intense stirring of the reactor vessels in order to facilitate the mass transfer. Oil conversion takes place within a few hours under normal conditions, particularly at temperatures between 50 and 70 °C (higher temperatures are necessary when ethanol is used instead of methanol). The glycerol is removed after the reaction is complete by allowing the two phases to form and settle. As shown in Figure 7.2, excess alcohol that did not react with the catalyst is separated from both phases (esters and glycerol) and recycled back into the reactor. Removing and recycling the catalyst and unreacted excess alcohol increases the complexity and operating costs of the process.

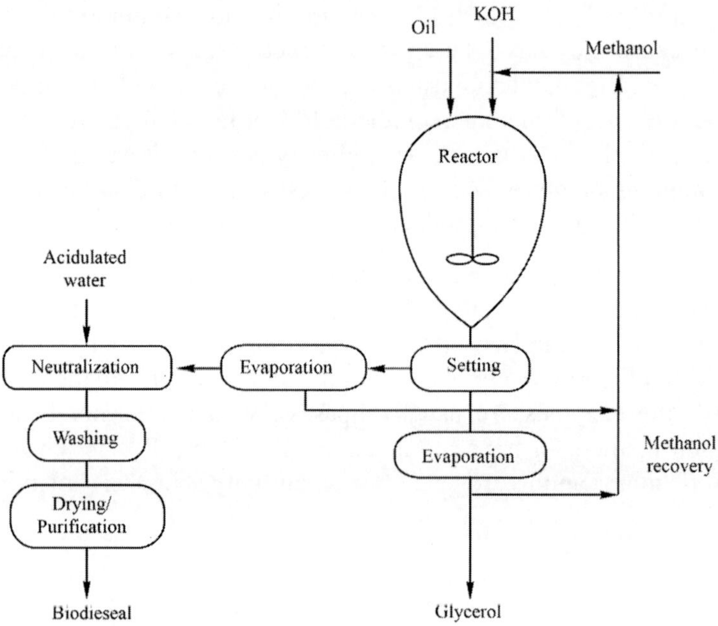

Fig. 7.2.: Schematic of biodiesel production process

7.5 Characteristics of Biodiesel

The properties of biodiesel as an automotive fuel are similar to those of petroleum diesel. Therefore, it can be directly installed in existing diesel engines without requiring major or minor modifications. The fuel can be used alone or in any ratio with petroleum diesel. In general, the most commonly used blend is B20, which is a mix of 20% biodiesel and 80% petroleum diesel. The composition of biodiesel is 11% oxygen by weight and essentially free of sulphur and aromatic compounds.

7.5.1 Physical Properties

Some of the physical characteristics of biodiesel are given in Table-8.1.

Table 8.1: Physical Properties of Biodiesel

S.No.	Properties	Values
1.	Specific gravity	0.88
2.	Viscosity @ 20 Ċ (centistokes)	7.5
3.	Cetane Index	49
4.	Cold filter Plugging Point (°C)	-12
5.	Net Heating Value (kJ/Liter)	33,300

7.5.2 Toxicity of Biodiesel

A significant factor in determining the suitability of biodiesel for commercial applications is its impact on human health. Several factors can affect health, including fuel toxicity and exhaust emissions. Wil Research Laboratories conducted a single-dose study on rats to examine the acute oral toxicity of pure biodiesel fuel and B20. The study concluded that biodiesel is not toxic, and there are no adverse effects expected from ingestion incidental to industrial exposure. Comparatively, the acute oral LD50 (lethal dose) is greater than 17.4 grams per kilogram of body weight, which is far less toxic than even table salt. Approximately 1000 mg/l of biodiesel is considered to be the lethal concentration for bluegills after 96 hours, according to the National Institute for Occupational Safety & Human Health. The following are other related effects of biodiesel:

- Human skin is very mildly irritated. Irritation caused by 4% soap and water solution is much less than that caused by 6% soap and water solution.
- The product is biodegradable. In comparison with conventional diesel fuel, it degrades at least four to five times faster.
- Biodiesel has a flash point of approximately 300 degrees Fahrenheit, which is much higher than conventional diesel fuel.
- Spills of biodiesel can decolorize any painted surface if left for an extended period of time.
- An automobile running on biodiesel (20% RSME + 80% diesel) does not exhibit a tendency for its exhaust gas mutagenicity to increase.

7.6 Prospective Feedstock in India

Various types of plants and oilseeds can be extracted as a source of oil, and there is a wide variety of these plants. It is important to take into account the fact that only plants that do not produce edible oil in significant quantities and that can be grown on wastelands on a large scale can be considered for biodiesel production in India. Aside from this, there are some plants and seeds that are considered to have the highest medicinal value. Therefore, they may not be suitable for the production of biodiesel in India. Due to the combination of all the above options, it appears that the following trees will likely be able to yield biodiesel in India:

(1). *Jatropha curcas* or Ratanjot

(2). *Pongamia pinnata* or Karanj

(3). *Calophyllum inophyllum* or Nagchampa

(4). *Hevea brasiliensis* or Rubber seeds

(5). *Calotropis gigantia* or Ark

(6). *Euphorbia tirucalli* or Sher; and

(7). *Boswellia ovalifololata.*

There is no doubt that Jatropha curcas is one of the most promising plants for the production of biodiesel among all the above-mentioned plants. In a rain fed environment, a Jatropha plantation of one hectare with 4400 plants can yield about 1500 liters of oil from one hectare. For a 10% replacement of petroleum with oil, about 3 million hectares of plantation are estimated to be necessary to replace it with oil. After the extraction of oil from Jatropha trees, the residue oil cake can be used as organic fertilizer after the oil has been extracted. In addition, one acre of Jatropha plantation could produce oil sufficient to meet the energy requirements of a family of five members. When used as fertilizer, the oil cake could be used to cover one acre.

7.7 Economic Feasibility

The cost of extracted oil in Indian conditions will be approximately Rs. 50-55 per liter. It is still noteworthy that despite these factors, the price of commercially available biodiesel is currently between Rs. 70 and 75 per liter. With the recent increase in the price of crude oil, it is imperative that we look for alternatives to fossil fuels. This will benefit both the economy of the country and the environment at the same time. In addition to providing better energy services at the village level as a result of the extraction of bio-fuels, the technology can also be widely spread throughout the area.

7.8 Biodiesel's Advantages

There are a number of advantages associated with using biodiesel as a replacement for diesel fuel, including:

- There are no emissions of sulfur dioxide (SO_2).
- Vegetable oils and animal fats can be used as renewable fuel.
- In comparison to diesel fuel, it has a low level of toxicity.
- May be blended with diesel fuel in any proportion; both fuels may be mixed during the fueling process.
- As biofuels degrade more rapidly than diesel fuel, spills resulting from their use are less likely to have adverse environmental consequences.
- It reduces emissions of contaminants such as carbon monoxide, particulate matter, polycyclic aromatic hydrocarbons, and aldehydes.
- As the only alternative fuel that can be used without modification in a conventional diesel engine, it is the only alternative fuel that is user-friendly
- Cooking oils and meat processing fat residues can be used as raw materials for raw materials.
- The emission of carcinogenic substances has been reduced, reducing the risk of health problems.
- It has a high flash point (at least 100°C).
- As a lubricant, it has excellent properties.

7.9 Disadvantages of Biodiesel

Biodiesel as a substitute for diesel fuel has the following disadvantages:

- In comparison with diesel fuel, it has a higher freezing point. There may be inconveniences associated with this in cold climates.
- Because biodiesel has a lower calorific value, there is a slight increase in fuel consumption.
- There is a slight increase in nitrous oxide (NOx) emissions compared to diesel fuel.
- In addition to dissolving sediments and other contaminants from diesel fuel, it also eliminates

- In addition, the biofuel flushes away the residue from the tanks and fuel lines into the engine, where it can cause problems in the injection systems and valves. As a result, it is recommended that tanks are cleaned prior to filling with biodiesel.
- It is not recommended to store biodiesel for a long period of time (over six months) due to its lower stability than diesel fuel.

When biodiesel is blended with diesel fuel, these disadvantages are significantly reduced.

Multiple Choice Questions

1. What chemical reaction makes biodiesel?

   a) Transesterification b) Sublimation

   c) Fermentation d) Polymerization

2. What is a byproduct of producing biodiesel?

   a) Glycerin b) Salt

   c) Methanol d) Polymer

3. What cannot be used to as a biodiesel feedstock?

   a) Grease b) Algae

   c) Wool d) Soyabeans

4. Up to what biodiesel blend is safely used in diesel engines without modifications?

   a) B80 b) B20

   c) B100 d) B60

5. What typical catalyst is used to make biodiesel?

   a) Base catalyst b) Acid catalyst

   c) Ultraviolet light d) Platinum

6. Pure biodiesel does not emit what pollutant?

   a) Particulate matter b) Sulfer Dioxides

   c) Carbon Monoxide d) Nitrogen Dioxide

7. Biodiesel does what at cold temperature

a) Melt b) Freezes

c) Gels d) Separates

8. Washing biodiesel removes what?

a) Acid b) Dirt

c) Soap d) Sulfur

9. What does the XX stand for in the biodiesel blend designation?

a) Percentage of biodiesel b) Energy rating

c) Quality d) Carbon footprint

10. Biodiesel is also call what?

a) Glycerin b) Methyl ester

c) Vegetable oil d) Alcohol

Answer

1	a	2	a	3	c	4	b	5	a	6	b	7	c	8	c	9	a	10	b

8

Biogas Technology

8.1 Introduction

Biogas is produced by the anaerobic digestion of biodegradable materials with the help of microorganisms. Biogas is also considered as renewable energy source because of is environment friendly nature. Conversion of cattle dung, kitchen wastes and agricultural wastes into biogas have the potential to minimize the burden on the fossil fuel and reduce the environmental pollution. Majority of developing countries specially their rural and suburban areas faces lack of access to the electricity and cooking gas. Biogas technology is emerging as a sustainable technology for waste treatment which could help to tackle the waste disposal issue in developing countries. Biogas and digested substrate are two main product of the anaerobic digestion. Biogas is a mixture of methane and carbon dioxide with traces of other gases such as H_2S, N_2, O_2 and H_2. Biogas, is considered as a clean and sustainable energy source, and can replace conventional energy sources that are responsible for ecological and environmental issues. Biogas generally contain main components methane (55–75%), carbon dioxide (25–45%), and traces of nitrogen (0–5%), hydrogen (0–1%), hydrogen sulphide (0–1%), and oxygen (0–2%). About 90% of the gas volume generated is consist of methane (CH_4) and carbon dioxide (CO_2). Methane is the most essential of its constituents, especially for the combustion process in automobile engines. The digested slurry has a good use as a bio-fertilizer and can be used as substitute for inorganic fertilizer. Biogas can be utilized for domestic and industrial application. Anaerobic digestion is a cost-effective and ecologically beneficial method for processing a wide range of organic wastes, including sewage sludge. Organic waste is biologically decomposed and transformed into a type of biogas and other energy-rich organic molecules in the absence of oxygen. The advantage of using biogas is that it can be stored and use on demand for power production unlike solar or wind. Countries like China and India lead the biogas development programme.

8.2 Anaerobic Digestion

Anaerobic digestion is a series of biological process which is carried out by several bacteria to break down organic material into biogas as shown in Figure 8.1. The process occurs in the absence of the free oxygen.

Biogas production mechanism

Biogas production takes place through four different stages.

i. Hydrolysis

Hydrolysis or liquefaction is initial step in anaerobic digestion of organic material. Organic biomass is consisted of complex material such as large organic polymers, protein, fats and carbohydrates. These are converted into soluble substance of smaller molecules such as amino acid, fatty acid and simple sugars by the fermentative bacteria. The governing chemical reaction is as below-

$$C_6H_{10}O_4 + 2H_2O = C_6H_{12}O_6 + 2H_2$$

ii. Acidogenesis

In acidogenesis process, biomass and organic products passed through hydrolysis process, further breaks down by acidogenic microorganism. This leads to forms varieties of organic acids, volatile fatty acids. The main output of acidogenesis is acetic (acetic acid, butyric acid, and propionic acids). H_2 and CO_2. Hydrogen producing acetogenic and hydrogen consuming methanogenic bacteria (acidogenic bacteria). Acetic acid is a major substrate for methanogenes as below.

$$C6H_{12}O_6 \rightarrow 3CH_3COOH$$

iii. Acetogenesis

Acetogenesis is third stage which is responsible for production of acetate by bacteria called as acetogenic bacteria. These bacteria form acetate and H_2 from heavy fatty acid. Therefore, this stage s called as dehydrogenation stage. Volatile fatty acids, produced in acidogenesis are broken by homoacetogenic bacterias Acetobacterium woodii and Clostridium aceticum to facilitate the production of acetic acid, carbon di-oxide and hydrogen.

$$C_6H_{12}O6 + 2H_2O \rightarrow 2CH_3COOH + 2CO_2 + 4H_2$$

$$CH_3CH_2OH + 2H_2O \rightarrow CH_3COO^- + 3H_2 + H^+$$

iv. Methanogenesis

Methanogenesis stage involves the formation of the methane where acetic acid and hydrogen conversion to methane and carbon dioxide with the help of methanogenic bacteria. Methanogens are anaerobic bacteria which survives in oxygen free environment and vulnerable to small quantity of oxygen. The reactions of methanogenesis are shown below:

$$CH_3COOH \rightarrow CH_4 + CO_2$$

$$2C_2H_5OH + CO_2 \rightarrow CH_4 + 2CH_3COOH$$

$$CO_2 + 4H_2 \rightarrow CH_4 + 2H_2O$$

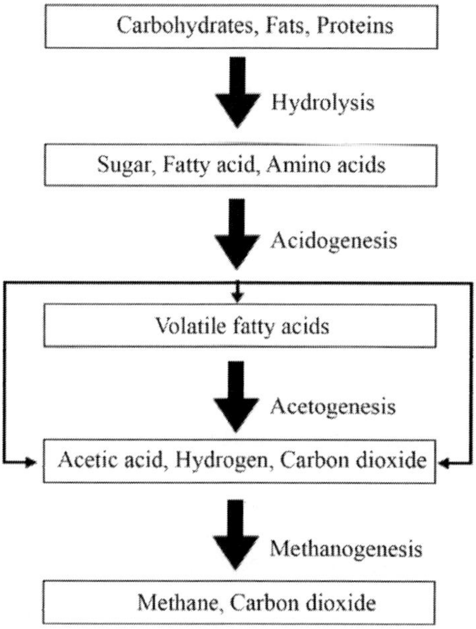

Fig. 8.1: Stages of anaerobic digestion

The following types of digestion are distinguished according to the temperature in the digester:

a) Psychrophilic digestion (10-20 °C, retention time over 100 days)

b) Mesophilic digestion (20-35 °C, retention time over 20 days)

c) Thermophilic digestion (50-60 °C, retention time over 8 days)

Generally, biogas contains methane and carbon dioxide gases as a major composition along with other gases as given in Table 8.1.

Table 8.1: Composition of biogas

Component	**Content [%]**
Methane, CH_4	50-75
Carbon dioxide, CO_2	25-50
Nitrogen, N_2	0-10
Hydrogen, H_2	0-1
Hydrogen sulphide, H_2S	0-3
Oxygen, O_2	0-2

8.3 Benefits of Anaerobic Digestion

i. Reduce odour and improve air quality

ii. Protect the water quality by reducing the potential pathogens associations with manure to enter surface and groundwater

iii. Generating energy heat or electricity for rural and urban areas

iv. Killing weed seeds in manure and control the weed in farms

v. Slurry as a bio fertilizer provide nutrients and reduce fertilizer cost in farm

vi. A biogas plant supplies energy and fertilizer.

vii. It improves hygiene and protects the environment.

8.4 Components of Biogas Plant

Biogas generation takes place in an air tight chamber called as biogas digester. The complete biogas system is consisting of different parts as below-

a. **Digester:** Anaerobic digestion process takes place in the biogas digester in the absence of air or oxygen. Biogas generated rises within he chambers and solid and liquid part settles at the bottom.

b. **Inlet:** Solid organic feed material homogeneously mixed and feed from the inlct

c. **Outlet:** The solid and liquid after complete digestion called as slurry rises through the slurry outlet

d. **Gas holder:** Biogas generated during the anaerobic digestion need to be collected and require a space which is generally provided the upper part of the digester. Floating drum biogas plant have the separate metal or fibre drum placed at the top of the digester. Fixed dome digester the upper part act as gas collector.

e. **Gas outlet:** Biogas is taken out through the gas outlet the carried the kitchen with the help of connection tube.

8.5 Types of Biogas Plants

Biogas plant are generally divided into two design

1. Floating drum type biogas plant
2. Fixed dome type biogas plant

The two models are fed daily as per the capacity of the plant.

8.5.1 Floating Drum Plants

Floating drum plants have main component as a digester and a gas holder for storing the biogas. Gas holder is in the floating stage because of the fermented slurry or a water jacket. Biogas is collected in the drum, and as the amount increases makes drum to rise in level and lowers down when gas is used.

KVIC biogas plant

The common example of this type biogas plant is KVIC (Khadi Village and Industrial Commission) model (Figure 8.2). KVIC model have underground constructed digester constructed from the bricks and cement with as a partition wall. A gas holder or drum made of mild steel is placed on the digester with wedge support and central guide frame. Inlet pipe for feeding the material and outlet for digested slurry are provided up to the bottom of the digester on either side of partition wall. The weight of the drum maintain he pressure on the gas to make the gas flow at the pressure through the outlet gas pipe. Metal component of the KVIC model increases overall cos of the biogas plant and accounts about 40% of the total installation cos of the biogas plant.

Advantages

- Simple to run and easy for construction

Drawbacks

- High cost of construction of the floating drum
- Metal can corrode reducing the overall lifespan
- Higher maintenance expense involved due to frequent painting

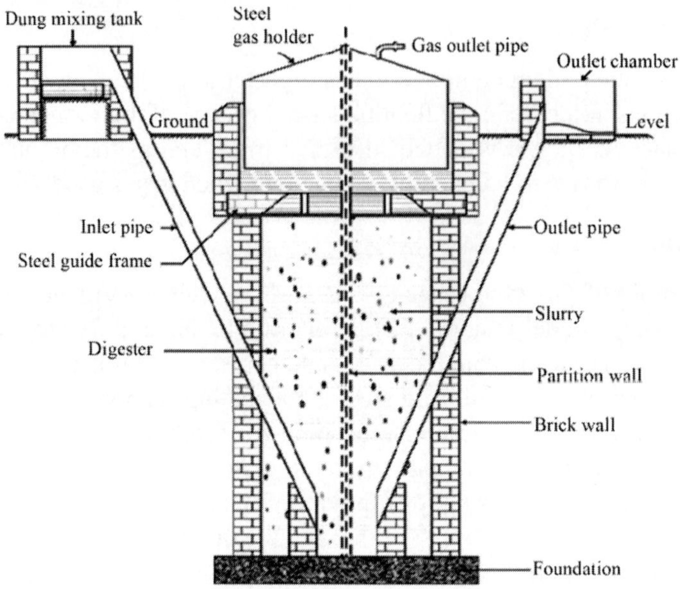

Fig. 8.2: KVIC model biogas plant

8.5.2 Fixed Dome Biogas Plant

A fixed dome plant is made up of a digester and a non-moving gas chamber. The biogas generated is stored in the upper part of the digester. When gas production is stopped, slurry is sent into the compensating tank. Gas pressure reduces as the gas amount get utilized.

Deenbandhu model biogas plant

Deenbandhu model biogas plant was developed by Action for Food Production (AFPRO), New Delhi, in 1984. The meaning of the world Deenbandhu is friend of the poor. The design (Figure 8.3) of this pant have two spherical parts joined together at their bases. The lower part act as a digester and upper part as a gas holder. Inlet and outlet tank are provided for feeding and slurry removal respectively. Outlet displacement chamber is provided above the normal slurry level of the outlet tank for accommodate he slurry displaced from the digester. The strength of brick masonry and cement allows to construct this type of biogas plant underground. The capacity of the Deenbandhu biogas plant recommended up to 6 m^3 per day.

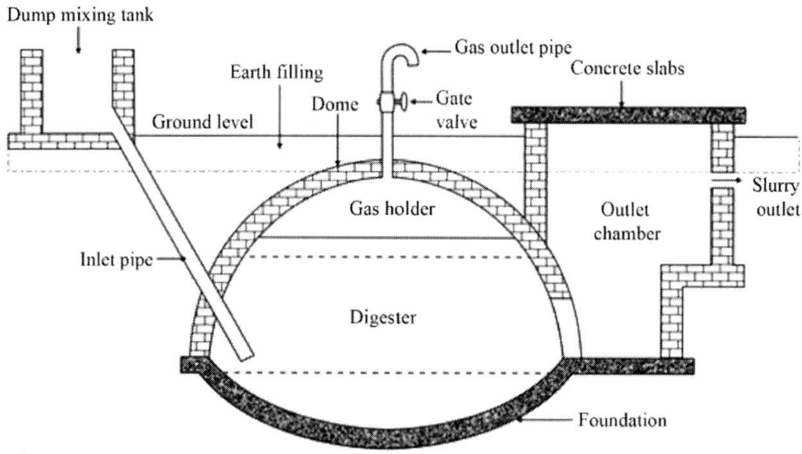

Fig. 8.3: Deenbandhu biogas plant

Janta model biogas plant

Janta Model Biogas Plant (Figure 8.4) developed by Gobar Gas Research Station, Ajitwal in 1978. The digester and the gas holder constructed using brick masonry structure. The digester is made of a shallow well having a dome-shaped roof on it. The inlet and outlet tanks are connected with the digester through large chutes which are called displacement chambers. The gas pipe is fitted on the dome and there is an opening on the outlet wall of the outlet displacement chamber for the discharge of spent digested slurry. The size of these plants is limited to 15 m^3 per day.

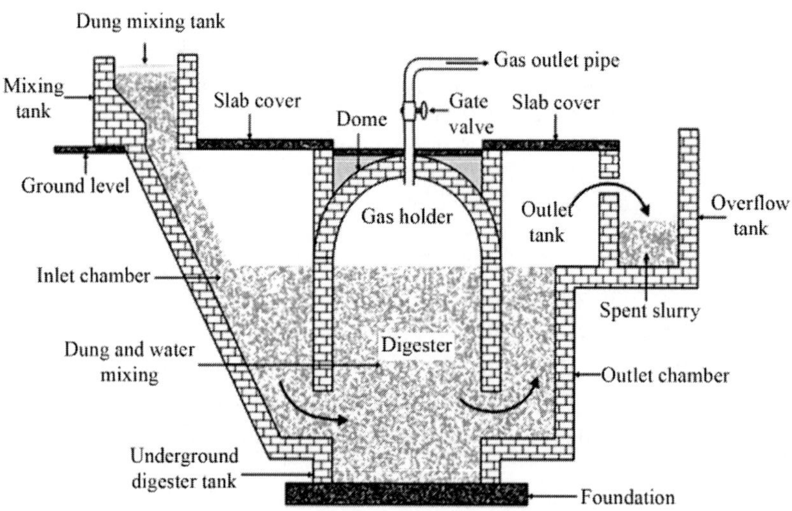

Fig. 8.4: Janata Biogas plant

Advantages

- Plant have lower cost and lack of moving components, which means it has a lower risk of air leakage and a longer lifespan

Disadvantage

- The porosity and cracks, drops the is a pressure drop of the gas.
- Require skilled supervising members or technicians

Anaerobic digesters further can be classified into

1) Standard rate (cold digester): It is a very simple type digester which can be used for a lengthy period of digestion of 30–60 days. It has four stages (a) scum layer (b) a liquid layer (c) a region of digesting solid (d) a region of digested solid.
2) Higher rate digester: It is a modified version of the generalized rate digester. In such digester, there is continuous feeding. The sludge is mixed and thickened. All the mixing creates a uniformity reducing the tank volume and casing a stable efficient process. There are heat exchangers included in this type of digester it is because of maintenance of constant temperature.
3) Two-stage digester: In this type of digester a secondary digester is attached with the high rate digester. The purpose of the secondary digester is to store the digested solid. Sometimes the secondary and the primary digester both have heat exchangers and are of similar design to serve as a standby digester.
4) Mesophilic and Thermophilic digestion: The high rate digesters are operated at a particular range of temperature. It is said to be a mesophilic digester if the temperature is around 30–38 °C. When the temperature is in the range of 50–57 °C, it is known as thermophilic.

8.6 Biogas Application

Biogas contains enough methane (CH4) to be utilised as a transportation fuel or for electrical or thermal application. Biogas, in the current scenario is being quickly gaining importance as a viable source of renewable energy. It can play an important part in environmental protection. There are several sources for obtaining biogas which may include industrial waste, urban solid waste, the organic fraction of urban solid waste, municipal wastewater, or industrial wastewater

8.7 Power Generation from Biogas

Power generation from urban, municipal and industrial waste

Sewage Sludge

Sewage sludge produces as result of wastewater treatment provided with chemical, physical and biological process before disposal. The sewage sludge contains mainly proteins, sugars, detergents, phenols, and lipids along with some toxic and hazardous organic and inorganic pollutants. Solid concentrations in these sludges can range from 0.25 percent to 12 percent. They are stabilised in either aerobic or anaerobic form before final disposal in regulated landfills, compost, or incineration process. Anaerobic stabilisation is recommended for sewage sludge as it can generate biogas. Biogas produced from wastewater can be utilised to generate electricity, heat, or cogeneration, as well as car fuel, or pumped into networks of natural gas distribution. The power generation from sewage based biogas generation (Figure 8.5) can be used power generation and treatment of sewage.

Table 8.2: Characteristic of sewage sludge

Characteristic	Values
pH	7.1–8.2
Bulk density (kg/L)	1.26–1.38
Particle density (kg/L)	2.40–2.56
Organic matter (g/kg)	418–592
Ash (g/kg)	345–440
Higher heating value (MJ/kg)	11.3–14.2
Organic carbon (g/kg)	205–403

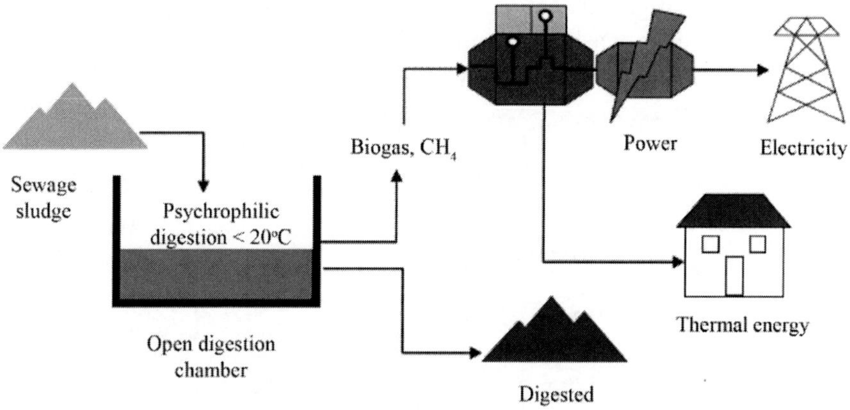

Fig. 8.5: Power generation process from sewage sludge

Organic Urban Waste

Urban organic wastes are an important component of urban and peri-urban agriculture. These wastes include the organic portion of municipal waste mainly street sweepings, street bins, household garbage, livestock manure and wastewater, households, and street sweepings, vegetable market waste, wastewater and livestock manure. Urban organic waste can be sent to bioreactors to produce biogas. The efficiency of the anaerobic digestion process can be increased by separating these wastes and diluting them in a fluid (wastewater) prior to treatment in an industrial anaerobic digestion reactor. In this case, the percentage of methane in the biogas can reach up to 75%.

Landfills

Landfills, are the sites where municipal solid waste (MSW) are disposed using suiable technology (Figure 8.6). Landfill use engineering principles to confine waste within the smallest possible area, i.e., minimizing its volume. Land fill gas (LFG) is a mixture of gases produced as a result of chemical and physical processes that take place within confined MSW due to the biodegradation of organic waste under anaerobic conditions. In a controlled landfill, hundreds of tons of solid waste treated with technical processes are added daily. The waste is placed on a geomembrane that prevents leachate from entering the ground; then, it is compacted and coated. In the middle of these layers of waste, conduction systems are connected to chimneys. As a result of the decomposition of organic matter and other wastes, a series of gases are produced that are emitted into the atmosphere. If the methane portion is sufficiently high for electricity production, it can be used in internal combustion engines as shown in Figure 8.7. To use these gases as vehicle fuel requires a cleaning process, as the waste has a high concentration of nitrogen. In the metropolitan cities of India, the disposal of waste in landfills would allow the recovery of 60% to 90% of biogas appropriate for energy production or for use as fuel.

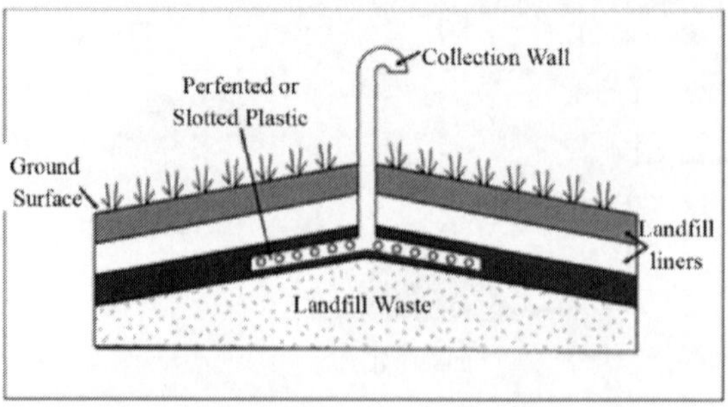

Fig. 8.6: Controlled landfill

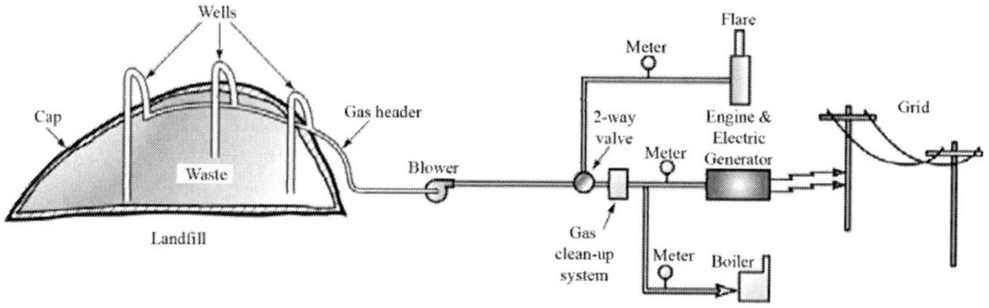

Fig. 8.7: Landfill gas recovery and power generation

8.8 Commercial Application of Biogas Plant

Biogas is produced on an industrial scale using the available different raw material through the use of different technologies. Both power (Figure 8.8) and thermal (Figure 8.9) application is possible using biogas. Digester is an important component of any capacity of the biogas plant. Family-sized biogas plants are mostly used to fulfil the demand of cooking and heating. Large-scale commercial biogas plants aim to meet energy needs for processing/ manufacturing or yield financial benefits by selling electricity and/or heat. Medium and large scale biogas plant are more beneficial than household size biogas plant. The cost per unit of gas is lower due to economic scale and use of advanced technologies. raw material as a feedstock for the large scale biogas plant can be available from various sources such as sewage sludge, food waste, livestock waste, municipal solid waste (MSW), agriculture waste and agro processing residues.

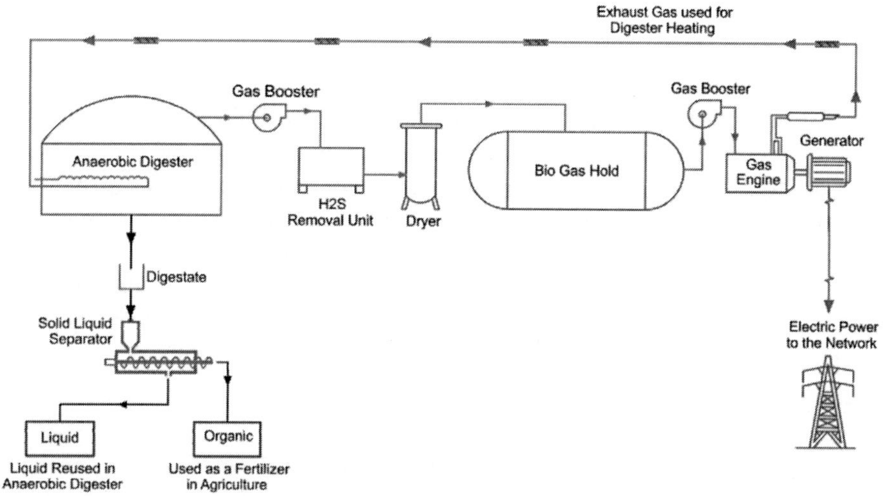

Fig. 8.8: Power generation application of biogas

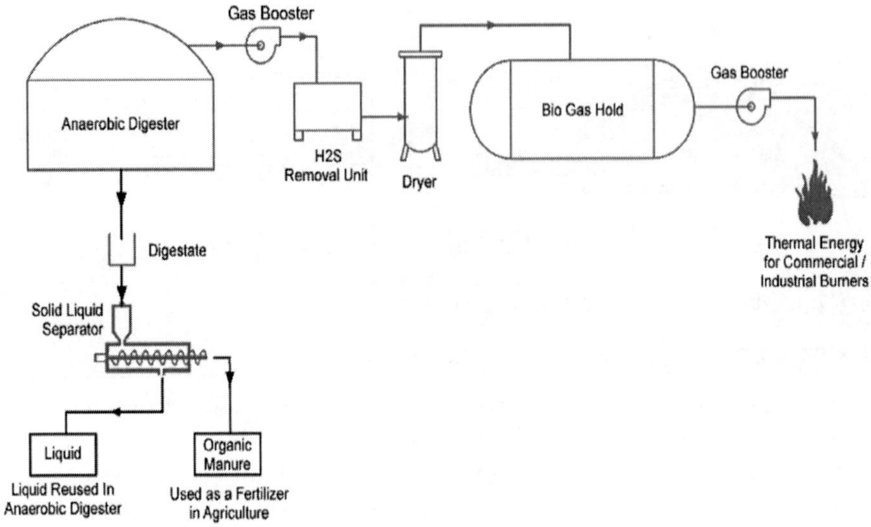

Fig. 8.9: Thermal application of biogas

Banas Bio-Gas Plant

Banas biogas plant is operated by Banas dairy, Asia's largest Co-operative Dairy established in 1969. It operates in Bansakantha District of Gujarat, Rajasthan and Uttar Pradesh. Banas Dairy have setup a pilot project Biogas generation plant at its Dama Semen station. A system is developed through which cow dung is collected from the villages which is collected from 254 farmers of 12 numbers of villages. The farmers are paid Rs.1 per kg of cow dung. The payment of cow dung is paid to farmer with the milk payment on every 15 days. Homogenous mixture (slurry) is prepared through agitator maintain the solid and water ratio in 1:1 ratio and then slurry is transferred to the anaerobic digester. The generated raw biogas is collected in the double membrane balloons. Digester top mounted balloon has 500 CUM capacity and raw gas storage balloon has 1000 CUM capacity. Raw biogas is purified for methane enrichment by removal of other gases and purified gas have methane content of more than 93%. The purified gas is then compressed to high pressure of about 200 Bar. Compressed purified gas is store in the cylinder cascade system. The purified gas is filled in the vehicle through dispensing system. It has quantity 800 kg of BioCNG per day. 100 vehicles can be filled through the gas station (8 kg per vehicle)

Table 8.3: Specification of biogas plant at Banas dairy

Capacity of digester	: 3000 m^3
Dung requirement	: 40 ton of cow dung
Raw Biogas	: ~2000 m^3/day
Bio-CNG	:~ 800 kg/day

Components of biogas plant

• Cow Dung Collection • Weigh Bridge • Unloading tank • Mixing Tank •Digester	• Gas Balloon • Purification System • Compression Process • Cascade System • Dispenser	• Solid Liquid Separator • Vermi-compost • Liquid Fertilizer Tank

Multiple Choice Questions

1. Which of the following is a substrate for biogas production?

   a) Municipal and residential waste b) E-waste

   c) Metallic waste d) Gaseous effluents

2. Which of the following is a preferred substrate for biogas production? Note that TS stands for total solids.

   a) Less than 1% TS b) 20-40% TS

   c) 1-5% TS d) 5-10% TS

3. Which among the following is the best source for methane production?

   a) Metallic scrap b) E-waste

   c) Plastic waste d) Water hyacinth

4. Which of the following is used to produce biogas from biomass?

   a) Anaerobic treatment b) Aerobic treatment

   c) Fermentation d) Pyrolysis

5. Which of the following best indicates the steps of anaerobic digestion?

   a) Waste water feed → biogas storage → generator → biogas

   b) Waste water feed → digester → biogas → biogas storage → generator

   c) Generator → waste water feed → digester → biogas → biogas storage

   d) Waste water feed → biogas → digester → biogas storage → generator

6. What occurs in the hydrolysis step of anaerobic digestion?

a) Large polymers combine with water molecules

b) Large polymers break down to form water molecules

c) Small polymers combine to form large polymers with the help of water molecules

d) Large polymers break down into amino acids, fatty acids and simple sugars

7. Which of the following is produced in acetogenesis?

a) Ethanol b) Acetate

c) Acetone d) Ketone

8. Which of the following is produced apart from acetates in acetogenesis step in anaerobic digestion?

a) Carbon monoxide b) Charcoal

c) Carbon dioxide d) Acetone

9.What are the two main products of anaerobic digestion?

a) Carbon monoxide and hydrogen

b) Methane and carbon dioxide

c) Methane and carbon monoxide

d) Hydrogen and carbon dioxide

10. Biogas contains

a) CO_2 b) H_2S

c) CH_4 d) All of these

11. The technology of biogas production in India was

a) Imported from Japan

b) Imported from Denmark

c) Imported from USA

d) Developed in India due to efforts of IARI and KVIC

12. Which of the following statements about methanogens is not correct?
    a) They can be used to produce biogas
    b) They are found in the remen of cattle and their excreta.
    c) They grow aerobically and breakdown cellulose-rich food
    d) They produce methane gas
13. Which types of plants are fed and emptied regularly?
    a) Batch type plants
    b) Continuous type plants
    c) Dome type plants
    d) Drum type plants
14. Indian fixed dome digesters are designed for holding what pressure capacity?
    a) 0 -90 cm of water column
    b) 70 -90 cm of water column
    c) 50 -65 cm of water column
    d) 10 -25 cm of water column

15.Khadi village industrial type biogas plant is an example of

a) Flexible bag biogas plant
b) Fixed dome biogas plant
c) Floating drum biogas plant
d) Semi-batch type

16.What statement is most true about anaerobic digester mixing?

a) Need to bring food (VS) into contact with bacteria
b) Provide infrequent mixing
c) If the mixing system isn't working, increase feed rate
d) Proper mixing is not that important to efficient digester operation

Answers

1	a	2	b	3		4	a	5	b	6	d	7	b	8	c	9	b	10	d
11	d	12	c	13	b	14	a	15	c	16	a								

9

Fuel Cell

9.1 Introduction

Hydrogen is colourless and odourless gas. Hydrogen is simplest and lightest element and most abundant available element on the earth. It is not primary energy source as it does not exist in free state however it presents in compound with oxygen (water) and carbon, nitrogen and oxygen in living material and fossil fuels (natural gas, coal and biomass). Therefore, hydrogen can be produced from these sources using some external energy. Hydrogen can be burned without generating CO_2 and have clean combustion. but with air it can form very explosive mixture. Hydrogen is considered as an energy carrier. Like energy carrier allows the transport of energy in usable form from one place to another. Hydrogen like electricity is an energy carrier that must be produced from another energy source. Hydrogen is high energy density fuel by weight however it contents lower energy by volume.

9.2 Application of Hydrogen

As an energy carrier, hydrogen has varied applications in such as stationary, mobile, backup and speciality. Stationary application can be in power plants or fuel cells, whereas mobile use consists of automobile, airplanes or boats. Back up use can be at home or office utility using portable generator.

9.3 Production of Hydrogen

Hydrogen can be produced form primary energy sources like fossil, nuclear or biomass using various process.

i. **Steam reforming:** Steam reforming process is used to produce hydrogen from fossil fuels. Hydrocarbon reacts with water steam to produce hydrogen. High temperature and pressure is used in the presence of catalyst.

ii. **Gasification of coal and biomass:** Gasification process converts carbonaceous material like coal or biomass into the syngas or producer gas which is composed of various mixed gases including hydrogen. The concentration of hydrogen is depending on various factor like moisture content in fuel use, catalyst and oxidation agents use.

iii. **Electrolysis of water:** Electrolysis is a process which used electrical energy to transform the chemical energy of the water into the gases. Using the direct electric (DC) current, redox reaction takes place which decomposes the water into hydrogen and oxygen. For proper reaction and conductivity, an electrolyte like NaSO4, NaOH or H2SO4 could be dissolved in water.

iv. **Thermolysis and Thermochemical water splitting:** These methods used thermal energy to produce hydrogen. Thermolysis involves water dissociation which is single step process at high temperature (>2500 K). thermochemical water splitting involves series of chemical reactions to dissociate the water into hydrogen and oxygen at temperature between 600 K to 1200 K.

v. **Photo-electrochemical water splitting:** Photo-electrochemical splitting of water is chemical reaction which used photon energy to split the water. Electromagnetic waves like ultraviolet light, X rays and gamma rays having short wave length with high energy can be used for the process.

9.4 Advantages and Limitations of Hydrogen Fuel

i. Hydrogen is renewable in nature, clean fuel and available in abundant quantity

ii. It is non-toxic which provide advantage over other fuels

iii. Fuel conversion efficiency using hydrogen is higher compared other fuels

Limitations

i. Hydrogen is highly explosive

ii. Presently it is expensive

iii. Storage of hydrogen is difficult as it requires very low temperature or in compressed state

iv. Transportation is difficult

v. It is odourless, so difficult to detect leakages and need special sensors

9.5 Fuel Cell Technology

First fuel cell was invented by William Grove in 19th century. In 1839, Grove demonstrated hydrogen-oxygen fuel cell. In 1939, Francis T Bacon used Pt catalyse in H2-O2 fuel cell and later patented in 1954. The first successful application of fuel cell was in space programme for electrical power and at the same time provide heat and potable water. NASA's apollo space programme used alkaline fuel cell system during 1960-1965.

9.6 Fuel Cell

Fuel cell is an electro-chemical device which converts chemical energy into electricity. It consists of two electrodes positive (cathode) and negative (anode). An electrolyte as a conducting material is used which carries the electrically charged particles form one electrode to another. The chemical reaction between the fuel and the oxidizer takes place at the electrode which produces the charged particles. Electrolyte acts as a medium for carrying these particles. The fuel generally used in fuel cell is hydrogen or source of the hydrogen like methanol and oxidizer may be oxygen or air. The major advantage of the fuel cell is that it produces no emission, as water in vapour form is by-product. The single fuel cell unit generated a small amount of current (DC) therefore for the higher capacity combining several fuel cell is required depending on the power requirement which is called as a stack

Working principle of fuel cell

The fuel cell works on electrochemical conversion principle to produce electrical energy. generally, hydrogen fuel or hydrogen atoms enter at anode whereas at the cathode electrode, oxygen is supplied as shown in Figure 9.1. The reaction of hydrogen at anode releases electrons and hydrogen atom gets ionized. The ionized positive charged protons pass through the electrolyte to the cathode whereas the negative charged electrons travel through the external circuit to produce the small amount of DC current. The hydrogen ions travelled through electrolyte at the cathode combines with electrons returning to the fuel cell completing its electrical circuit and oxygen (or air) to produce the water vapours. In certain types of fuel cell, oxygen picks up the electrons at cathode, travel to the anode and combines with hydrogen ions to produce the water vapour.

The reaction at anode

$2H_2 \rightarrow 4H^+ + 4e^-$

At cathode, oxygen reacts with electrons and H^+ ions to from water

$O_2 + 4e^- + 4H \rightarrow 2H_2O$

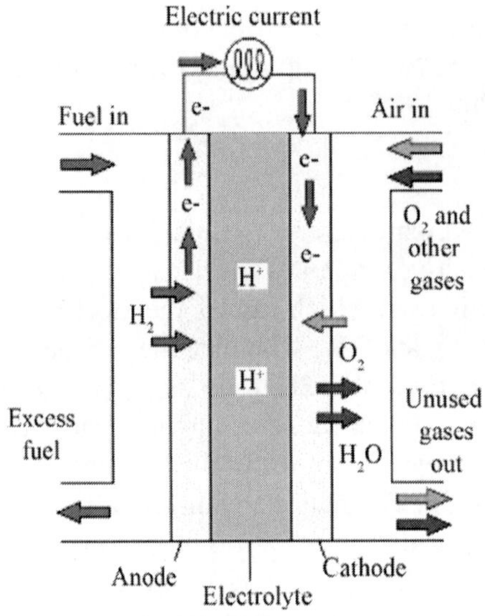

Fig. 9.1: Fuel cell

The fuel cell type is determined on the basis of which type of electrolyte is used. Electrolyte plays an important role in fuel cell operation as it allows only ions to pass between the electrode while electrons are not allowed to pass through the electrolyte which make electrons to pass through eternal circuit. As along as the hydrogen fuel is supplied, the electricity will be generated by the fuel cell.

9.7 Classification of Fuel Cells

Fuel cells are classified on the basis of factors like type of fuel used, operating temperature and the most common on the basis of electrolyte used. The different types of fuel cells are –

1. Alkaline fuel cell
2. Phosphoric acid fuel cell
3. Proton exchange membrane fuel cell
4. Solid polymer fuel cell
5. Molten carbonate fuel cell
6. Solid oxide fuel cell
7. Microbial fuel cell

Alkaline Fuel Cell

Alkaline fuel cell (AFC) uses aqueous alkaline electrolyte and operates a 60-80 °C (Figure 9.2). At lower temperature, generally catalysts with high activity are used which may be group of the platinum, silver or nickel. However, nickel is mostly used catalyst. The alkaline fuel cell are intolerant to the carbon dioxide and therefore require to remove the CO2 from the air. The fuel used in the operation of fuel cell is alcohol or hydrocarbon. The electrode material may be palladium plus silver as help to prevent the CO2 into the electrolyte.

The chemical reaction involves are

Anode $H_2 + 2OH^- \rightarrow 2H_2O + 2e^-$

Cathode $\frac{1}{2}O_2 + H_2O + 2e^- \rightarrow 2OH^-$

Overall reaction $H_2 + \frac{1}{2}O_2 \rightarrow H_2O$

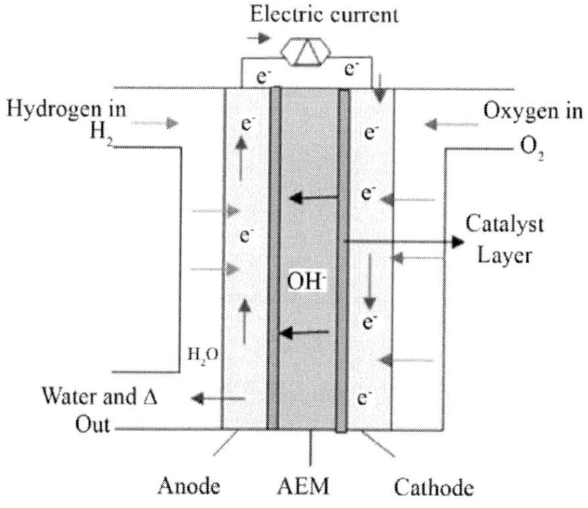

Fig. 9.2: Alkaline Fuel Cell

Phosphoric Acid Fuel Cell

Phosphoric acid fuel acid (PAFC) uses phosphoric acid as an electrolyte. Acids are generally tolerant to CO_2 and allow to use normal air and non-pure hydrogen (Figure 9.3). The electrode are made of of gold, tatalum, titanium and carbon. The catalyst used is platinum supported on graphite. The operating temperature is about 200 °C. Acid fuel cells give lower efficincies than alkaline fuel cell due to poor performance of air electrode.

Anode reaction: $H_2 \rightarrow 2H+ + 2e-$

Cathode reaction: $½O_2 +2H+ + 2e- \rightarrow H_2O$

Overall reaction: $H_2 + ½O_2 \rightarrow H_2O$

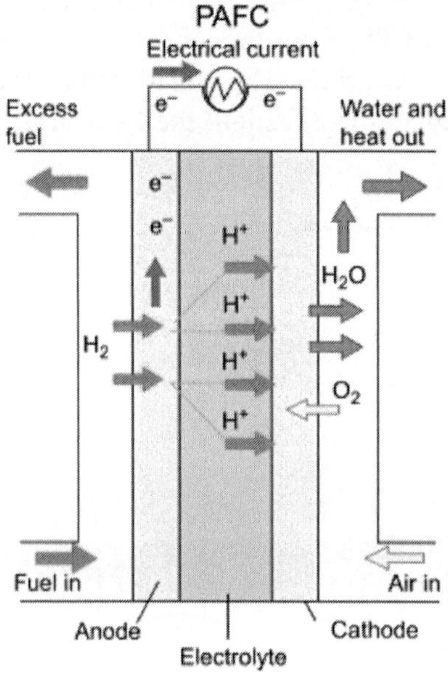

Fig. 9.3: Phosphoric Acid Fuel Cell

Proton exchange membrane fuel cell

The electrolyte used in proton exchange membrane fuel cell (Figure 9.4)is a layer of solid polymer which has ability to pass the protons. The operation takes place at lower temperature at about 90 °C and uses hydrogen and oxygen as fuel and oxidant respectively. The cell requires cooling arrangement and exhaust water management for functioning. Solid electrolyte provides the safety for transportation application, suitable for small capacity localized power generation. Heat can be used for hot water or space heating.

Anode reaction: $H_2 \rightarrow 2H+ + 2e-$

Cathode reaction: $½O_2 +2H+ + 2e- \rightarrow H_2O$

Overall reaction: $H_2 + ½O_2 \rightarrow H_2O$

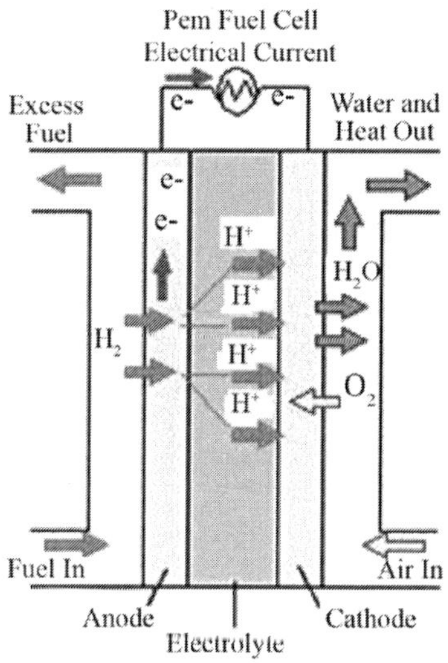

Fig. 9.4: Proton exchange membrane fuel cell

Solid polymer fuel cell (Methanol fuel Cell)

Solid polymer fuel cell uses methanol as fuel and based on solid polymer technology which is early stage of development. This reduced the obstacle in storage or generation of hydrogen in automotive area. The issues like lower electrochemical activity, tendency of methanol mixing with water which causes to pass the water saturated membrane and causing corrosion.

Anode reaction $MeOH + H_2O \rightarrow CO_2 + 6H+ + 6e-$

Cathode reaction $3(\frac{1}{2}O_2) + 6H+ + 6e- \rightarrow 3H_2O$

Overall reaction $MeOH + H_2O + 3(\frac{1}{2}O_2) \rightarrow CO_2 + 3H_2O$

Molten carbonate fuel cell

Molten carbonate fuel cell (Figure 9.5) uses molten mixture of potassium carbonate and lithium carbonate as electrolyte to transport carbonate ions from the cathode to the anode. The operating temperature ranges about 850 °C. The catalyst used is nickel.

the reaction process occurs in hydrogen oxygen fuel cell at air electrode

$O_2 + 4e- \rightarrow 2O_2^-$

When it moves to the fuel electrode to oxide the hydrogen

$H_2 + O_2 \rightarrow H_2O + 2e^-$

Molten ionic oxide has melting point higher than 1000 °C. Carbon dioxide produced can have the issue with some salts

$CO_2 + SO_4^{2-} \rightarrow CO_3^{2-} + SO_3$

Therefore, using the molten carbonate or mixture of carbonates gives the satisfactory results.

Anode reaction: $H_2 + CO_3^{2-} \rightarrow CO_2 + H_2O + 2e-$

Cathode reaction: $½O_2 + CO_2 + 2e- \rightarrow CO_3^{2-}$

Overall reaction: $H_2 + ½O_2 \rightarrow H_2O$

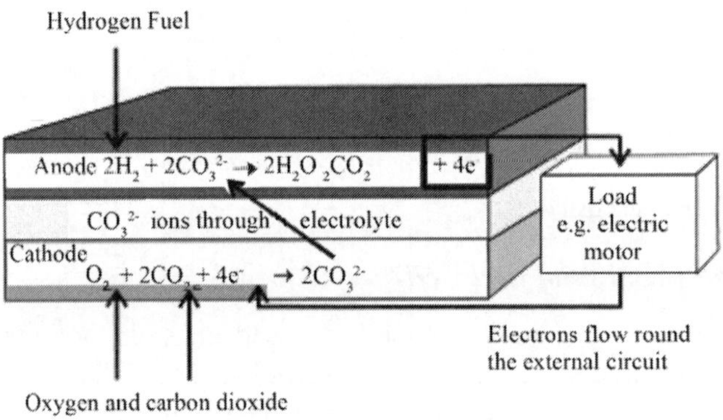

Fig. 9.5: Molten carbonate fuel cell

Solid oxide fuel cell (SOFC)

Solid oxide fuel cell (Figure 9.6) use ion conducting oxide ceramic as electrolyte and operates in temperature 900-1000 °C. SOFC provides advantages like lower operating issues, higher efficiency (50-60%), combined heat and power application and internal reforming of the hydrocarbon fuel.

Anode reactions: $H_2 + O^{2-} \rightarrow H_2O +2e-$

$CO + O^2- \rightarrow CO_2 + 2e-$

Cathode reaction: $O_2 + 4e- \rightarrow 2O^{2-}$

Overall reaction: $H_2 + O_2 + CO \rightarrow H_2O + CO_2$

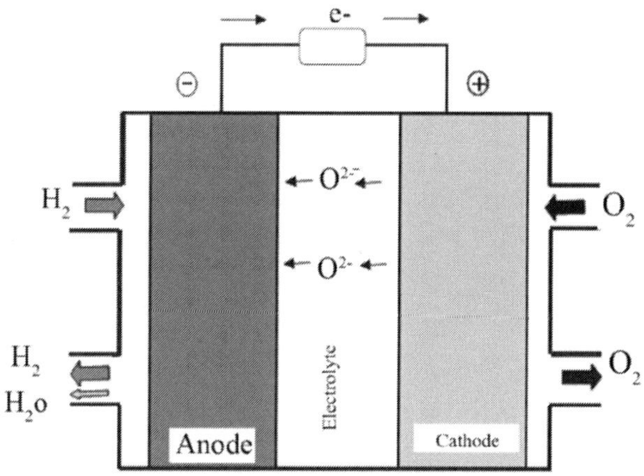

Fig. 9.6: Solid oxide fuel cell (SOFC)

Microbial fuel cell

Microbial fuel cell uses active microorganism as a catalyst (Bio-catalyst) in a anaerobic compartment which is considered as anode. As shown in Figure 9.7, it has separate anode and cathode chamber which are separated physically by proton exchange membrane. The organic substrate in the anode compartment is oxidized by the active biocatalyst to producer the protons and electrons. The protons are carried to the cathode chamber through PEM whereas electrons passes through the external circuit. Protons and electrons reacts in the cathode chamber, and at the same time reduction of oxygen to water takes place.

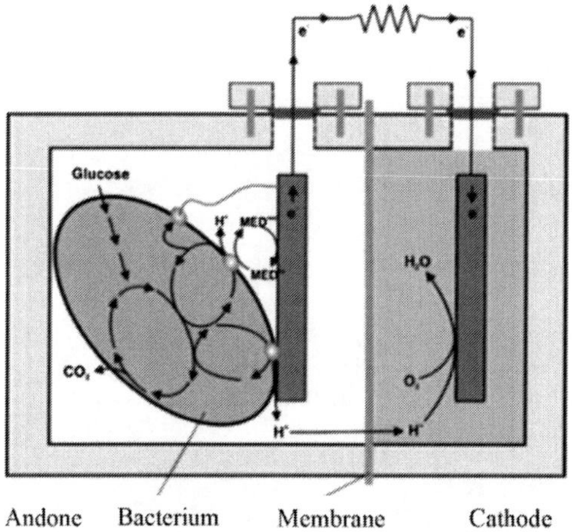

Fig. 9.7: Microbial fuel cell

9.8 Application of Fuel Cell

A. **Portable Power**

i. Laptops
ii. Cellular phones
iii. Power tools
iv. Military equipment
v. Battery charger
vi. Unattended sensors
vii. Unmanned aerial
viii. Underwater vehicles

B. **Backup power**

i. Computer systems
ii. Manufacturing facilities
iii. Homes
iv. Utility substation

C. **Transportation application**

i. Automobiles
ii. Buses
iii. Utility vehicles
iv. Scooters
v. Bicycles

D. **Stationary power application**

i. Primary power source for homes
ii. Hybrid power system connected to photovoltaics, batteries, capacitors or wind turbines

9.9 Fuel Cell Thermodynamics

Fuel cell efficiency

The input energy and output energy of the fuel cell can be described in following Figure 9.8.

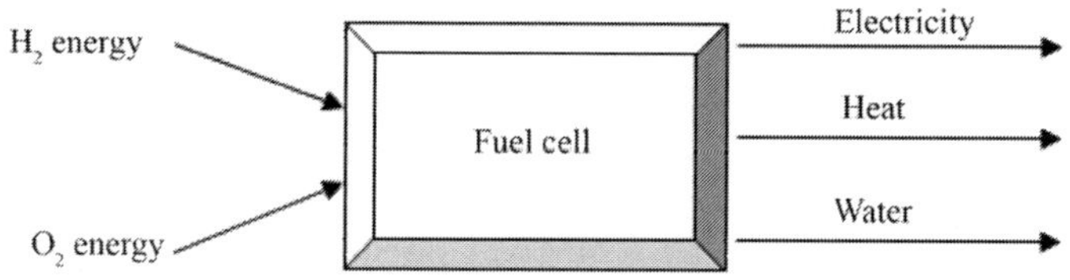

Fig. 9.8: Energy input and output in fuel cell

The electrical power and energy of the fuel cell can be calculated from the common formula as-

Power = V x I

Energy = V x I x t

The terms like enthalpy, Helmholtz and Gibbs free energy are used to define the chemical energy in the fuel cell.

Gibbs free energy is defined as the energy available to do external work, neglecting the work done by changes in pressure and volume. The external work in fuel cell involves moving electrons round the external circuit, whereas any work done by change in volume between the input and output is not harnessed by the fuel cell. Energy is all external work that can be extracted including the work done by change in pressure and volume. Enthalpy is the sum of Gibbs free energy and energy associated with the entropy.

$G = H - TS$

Where

G = Gibbs free energy

H = Enthalpy

T = Temperature

S = Entropy

The energy is released by fuel cell is defined by the change in Gibbs energy before and after the reaction which can be represented as –

$\Delta G = G_{output} - G_{input}$

The change in Gibbs free energy is not constant and changes according to the temperature and state of the material. Table 1 shows the Gibbs free energy for the basic hydrogen fuel cell reaction for the reaction

$$H_2 + ½\, O_2 \rightarrow H_2O$$

Table 9.1: ΔG values for water for different

Water form	Temperature °C	ΔG, kJ/mole
Liquid	25	- 237.2
Liquid	80	-228.2
Gas	80	-227.1
Gas	100	-225.2
Gas	200	-220.4
Gas	400	-210.3
Gas	600	-199.6
Gas	800	-188.6
Gas	1000	-177.4

If is considered that the process is reversible and no losses are considered, the Gibbs free energy is completely converted into the electrical energy. then efficiency of the fuel cell is given by –

$$\text{Efficiency}\left(\eta_{fc}\right) = \frac{\text{Electrical energy produced}}{\text{Change in Gibbs free energy}}$$

The energy released during the reaction process is precisely given by enthalpy of formation and mentions as negative as energy is released. Therefore, efficiency of fuel cell is given by

$$\text{Efficiency}\left(\eta_{fc}\right) = \frac{\text{Electrical energy produced / mole of fuel}}{-\Delta H}$$

The maximum electrical energy available is equal to the change in Gibbs free energy, therefore

$$\text{Maximum efficiency}\left(\eta_{max}\right), \% = \frac{\Delta G}{\Delta H} \times 100$$

10

Micro Hydropower

10.1 Introduction

As one of the oldest and most widely used renewable energy sources, hydropower generates electricity by using the natural flow of flowing water. Streams and rivers have been used for producing mechanical energy for thousands of years. Hydroelectric power plants are located near or on water sources since they derive their power from flowing water. The amount of energy that can be released in moving water depends on the volume of the water flowing and the elevation change from one point to another, often called head. Generally, hydropower plants can produce more electricity when there is more water flow and a higher head. Using water to drive turbines is the basic principle of hydropower. Dams and reservoirs are commonly used in hydropower plants, but they can also be used without them. The storage capacity of a hydropower dam with a large reservoir allows it to meet peak demand for short as well as long periods of time. There are also smaller dams that can be used both for pumping and generating electricity, such as those for night or day use, seasonal storage, and pumped-storage reversible plants. It involves producing hydropower on a smaller scale than dams and reservoirs, usually from a river-based facility that does not interfere with the flow of the river. As a result, it is considered that small-scale hydro to be more environmentally friendly than large-scale hydro.

10.2 Classification of Small Hydro Project

Small hydro projects (SHPs) harness the power of moving water, which drives a turbine to generate electricity up to 25 MW. Small hydropower projects are classified as micro/pico (with capacities up to 100 kW), mini (with capacities 101 kW-2 MW), and small (with capacities 2 MW- 25 MW) as presented in Table 10.1. As the turbine runs, mechanical energy is transformed into electrical energy.

Micro hydro power (MHP) plants are small hydroelectric power plants that produce up to 100 KW of electricity utilizing a flow of steam or water. It is sometimes possible to connect such systems to the public grid in order to supply electricity to remote homes or communities. Micro hydro systems are generally used to provide electricity generally in developing countries where there is no electricity grid available in rural villages or isolated communities. When electricity production is in surplus, surplus electricity can also be fed back into the national grid. A micro hydro scheme can be designed on a household basis or at a village level, utilizing local materials and labor. Most of the micro hydro plants in the developing world are in mountainous regions, such as in the Himalayas and Nepal, which has around 2,000 schemes, including mechanical and electric power plants.

Table 10.1: Classification of hydropower plant

Type	**Description**
Large Hydro	All installations with an installed capacity of more than 1000 kW (according to some definitions more than 10,000 kW)
Medium Hydro	Installations of 15 - 100 MW (usually feeding the grid)
Small Hydro	Installations of 1-15 MW (usually feeding into the grid)
Mini Hydro	Capacity between 100 - 500 kW (either as stand-alone schemes or more often feeding into the grid)
Micro Hydro	Installations with power output of 5 - 100 kW (usually provided power for small community or rural industry in remote areas away from the grid)
Pico Hydro	From a few hundred watts up to 5 kW

Advantages of micro-hydro power plant

a. **Efficient energy source**: Micro hydro uses a small amount of flow of water to generate electricity, and the power generated can be used up to a mile away from the plant.

b. **Reliable electricity source:** When compared to other small-scale renewable energy technologies, hydropower plants provide a constant and continuous supply of electricity. Peak energy seasons require large quantities of electricity.

c. **No reservoir required:** When the water passes through the turbines, it is diverted back into the river/stream, so that micro hydro systems have a relatively low impact on the environment.

d. **Cost effective energy solution**: Electricity requirements and location can determine the cost of a small hydro-power system. Also, the costs of operation and maintenance are relatively low.

e. **Power for developing countries:** Micro hydro power can provide electricity to small villages and communities in developing countries because it is low-cost, versatile, and longer operation.

f. **Integrate with the local power grid:** If there is a surplus of electricity, some companies can purchase it and integrate it into the grid.

g. **Environmental impact:** In comparison with fossil fuel-based power stations, the environmental impact is minimized. It is renewable and environmental friendly energy source

Disadvantages of micro-hydro power plant

a. Requirement of large initial capital

b. Suitable site characteristics are required

c. Energy expansion not possible

d. The summer months are likely to have less flow, resulting in lower power output, because streams will reduce their flow rates based on the season.

e. Hydro systems with small scales have a low ecological impact however, they must be considered before construction begins

10.3 Determination of the Power Output

The general formula used for the calculation of power from micro-hydro power is given as –

$P = \eta\, g\, Q\, H$

Where,

P = mechanical power produced at the turbine shaft, watts

η = hydraulic efficiency of the turbine

g = acceleration due to gravity (9.81 m/s^2)

Q = Flow rate passing through the turbine, litre/second

H = Head of water across the turbine, m

10.4 Use of Micro-hydro Power Plant

Power produced from a small hydro station can be used for various purposes, some of the uses have been classified as follows (Table 10.2):

1. **Productive Use:** Electricity is produced to perform activities that are used for services in exchange for money. Small businesses are most likely to experience these scenarios.

2. **Consumptive Use:** Electricity at the household or near the household is used for all the other purposes is consumptive use

Furthermore, power can be classified as mechanical or electrical based on its use, as well as its productive or consumptive nature

Table 10.2: Application of micro-hydro power plant

Application	Mechanical	Electricity
Productive use	• Agro-processing • Timber sawing • Textile fabrication • Cooling • Drying	• Mechanical uses with electricity as intermediate • Heating • Lighting • Fertiliser production
Consumptive use		• Domestic lighting • Cooking • Cooling • Radio and television

10.5 Working of Micro-hydropower System

Run-of-the-river type microhydropower systems consist of these basic components

a. Water conveyance channel, pipeline, or pressurized pipeline (penstock) that delivers the water

b. Turbine, pump, or waterwheel which transforms the energy of flowing water into rotational energy

c. Alternator or generator transforms the rotational energy into electricity

d. The regulator controls the generator

e. Wiring delivers the electricity.

The following illustration (Figure 10.1) shows the setup of a micro-hydro system. For water diversion the river water level has to be raised by a barrier, the weir. The water is diverted at the intake and conveyed by the channel along the landscape′s contour lines. The spillways protect against damage from excessive water flow. Water is slowed down and collected in the fore bay. The water then enters the penstock, where it is transported via pressure pipes to the power house, which houses the power conversion turbine, mill or generator. The turbine is the core of a micro-hydro power plant, which is driven by moving water. Different types of turbines are used depending on the head and the flow of the site. These turbines are used to rotate a shaft which is then used to drive the generator. The water is then discharged via the draft tube or a tail race channel in case of cross flow or Pelton turbines. In micro-hydro schemes, electricity is distributed remotely through a local grid. There must be a match between the demand output and the generator's capacity, otherwise the voltage and frequency can fluctuate, resulting in damage to some electrical equipment. Micro-hydro systems must vary their supply to keep tight control because the power demand is often variable since people switch lights and machines on and off. A water flow controller or an electronic load controller can be used to do this. The low-voltage direct current (DC) electricity produced by the system can also be converted to alternating current (AC) electricity using an inverter. Batteries are sometimes used to store electricity generated by standalone systems. Due to the seasonal nature of hydropower resources, batteries may not always be practical for micro-hydropower systems.

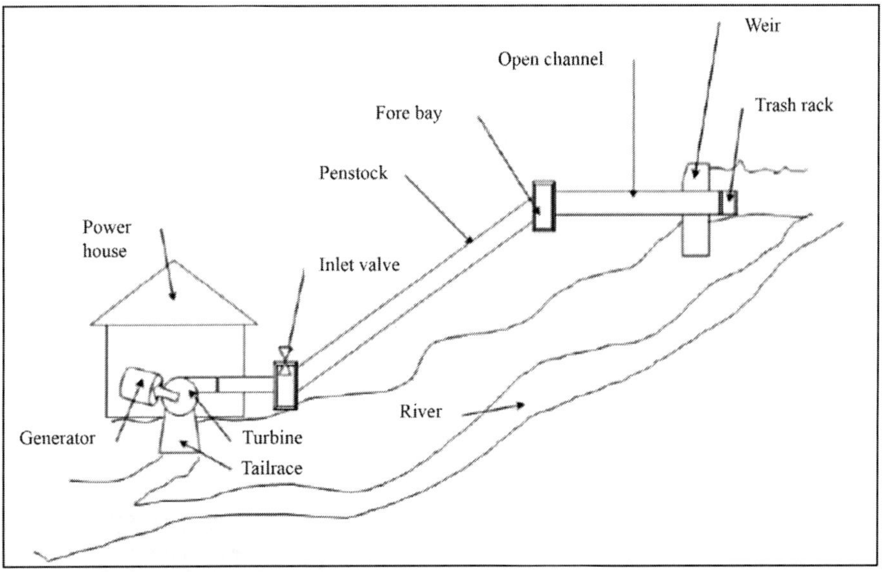

Fig. 10.1: Working of small hydro power plant

10.6 Types of Turbines Used

1. Impulse Turbines

In microhydro systems with high heads, impulse turbines are most commonly used. Turbine wheels are moved by water velocity, which is called the runner. Turgo and Pelton wheels are among the most common impulse turbine types.

Pelton wheel

A pelton wheel (Figure 10.2) generates energy by using jet force. A narrow nozzle is located at one end of a pressurized pipeline that ejects water. The water sprays out of the nozzle in a jet, striking the double-cupped buckets attached to the wheel. The impact of the jet spray on the curved buckets creates a force that rotates the wheel at high efficiency rates of 70–90%. Pelton wheel turbines are available in various sizes and operate best under low-flow and high-head conditions.

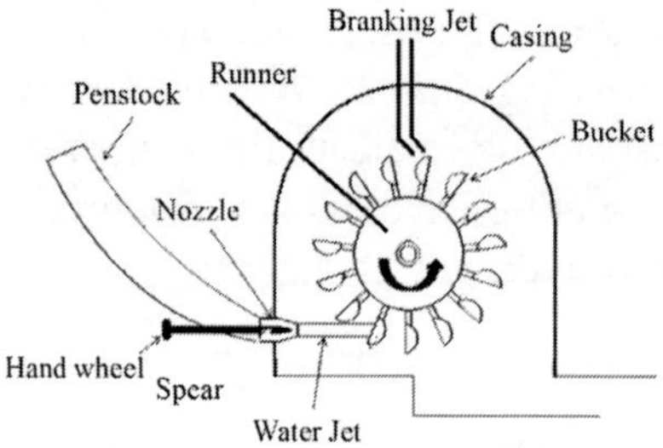

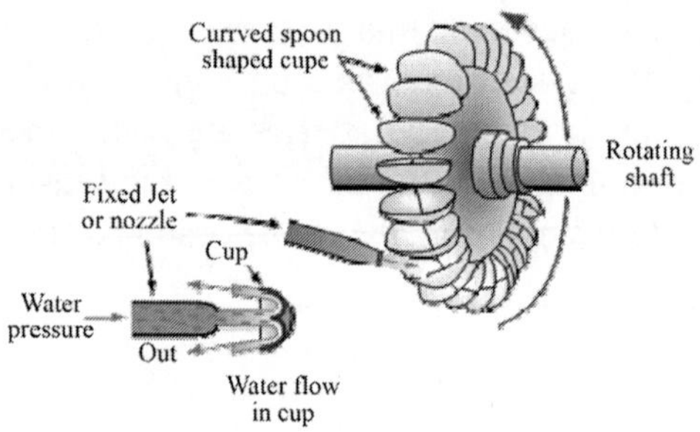

Fig. 10.2: Pelton wheel turbine

Turgo impulse wheel

The turgo jet is half the size of the pelton and is angled so that it sprays three buckets simultaneously as shown in Figure 10.3. Due to this, the Turgo wheel moves twice as fast. Additionally, it is less bulky, requires few or no gears, and is trouble-free to operate. Low-flow conditions can be accommodated by the Turgo, but medium or high head is required.

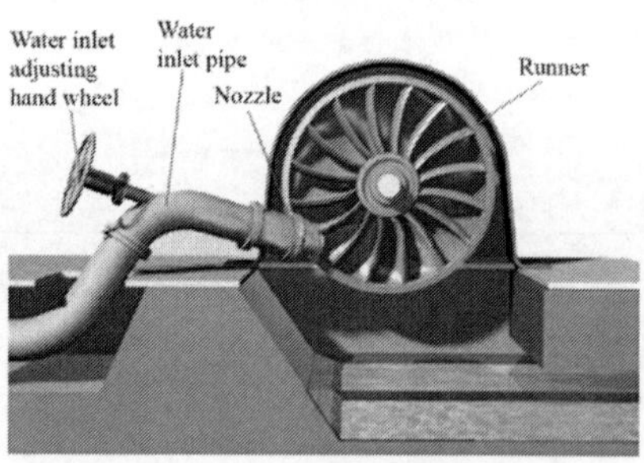

Fig. 10.3: Turgo impulse turbine

2. Reaction Turbines

A reaction turbine (Figure 10.4) produces energy by converting pressure instead of velocity. Reaction turbine blades remain in constant contact with water. Large-scale hydropower plants commonly use these turbines. A reaction turbine is not often used for microhydropower projects because of its complexity and high cost. A propeller turbine is an exception, since it comes in many different designs and works similarly to a boat's propeller. Usually, a propeller turbine consists of three to six fixed blades that are set at different angles and aligned along the runner. A propeller turbine is a bulb, a tubular turbine, and a Kaplan tubular turbine. In microhydro sites, Kaplan turbines can be used, which are highly adaptable propeller systems.

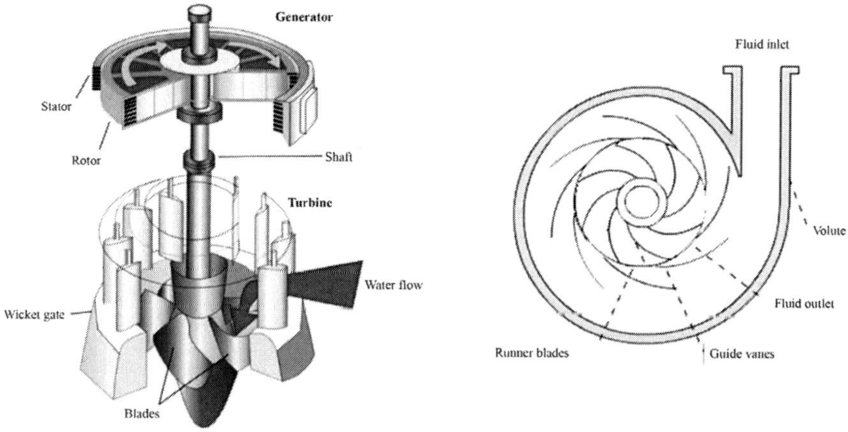

Fig. 10.4: Reaction turbine

Multiple Choice Questions

1. Which of the following is not a requirement for site selection of hydroelectric power plant?

   a) Availability of water b) Large catchment area

   c) Rocky land d) Sedimentation

2. The amount of electrical energy that can be generated by a hydroelectric power plant depends upon ___________

   a) Head of water b) Quantity of water

   c) Specific weight of water d) Efficiency of Alternator

3. Hydroelectric power plant is mainly located in ___________

   a) Flat areas b) Deserts

   c) Hilly areas d) Deltas

4. Which of the following is not an advantage of hydroelectric power plant?
   a) no fuel requirement
   b) low running cost
   c) continuous power source
   d) no standby losses
5. Which statement about hydroelectric power plant is wrong?
   a) Efficiency of hydroelectric power plant does not reduce with age
   b) Its construction coast is very high and takes a long time for erection.
   c) It is very neat and clean plant because no smoke or ash is produced.
   d) Meeting rapidly changing load demands is not possible in hydroelectric power plant.
6. Which of the following statement is true about hydroelectric power plant?
   a) Hydroelectric power plants are multipurpose.
   b) Due to non-uniform flow of water frequency control in such plants is very difficult.
   c) Hydroelectric power plant has high running cost
   d) Water is used as fuel in hydroelectric power plant
7. Dam having very wide base as compared to its height is called _________
   a) Buttress dam
   b) Arch dam
   c) Earth dam
   d) Solid gravity dam
8. Reaction turbine are employed for:
   a) Low heads
   b) Medium heads
   c) Both of these
   d) None of these
9. Impulse turbine are used for:
   a) Low heads
   b) Medium heads
   c) Both of these
   d) None of these
10. The power output from an hydroelectricity plants depends on
    a) Type of turbine, type of dam and type of catchment area.
    b) Type of dam, head, system efficiency.
    c) Type of dam, type of catchment area and discharge.
    d) Discharge, system efficiency.

11. In hydroelectric power plants
    a) Both operating and initial cost are high.
    b) Both operating and initial cost are low.
    c) Operating cost is low and initial cost is high.
    d) Operating cost is high and initial cost is low.
12. Water hammer is developed in a
    a) Penstock
    b) Draft tube
    c) Surge tank
    d) Turbine
13. A penstock is used as a condition between
    a) The turbine and discharge
    b) The dam and turbine in a hydro station
    c) The heat ex-change and the turbine in nuclear power plant
    d) None of above
14. Gross head of a hydroelectric power station is
    a) The height of water level in the river where the storage is provided.
    b) The difference of water level between the level in storage and tail race
    c) The height of water level in the river where the tails race is provided.
    d) All of above.

Answer

1	d	2	b	3	c	4	c	5	d	6	a	7	c
8	c	9	d	10	d	11	c	12	a	13	b	14	b

11

Ocean Thermal Energy Conversion

11.1 Introduction

More than 70% of Earth's surface is covered by the ocean, the world's largest solar energy collector and storage system. By utilizing only a small portion of this energy, we can meet all the energy needs around the world. As long as the sun shines and there are natural ocean currents, Ocean Thermal Energy Conversion (OTEC) energy source is free, abundant, and continually replenished. It is estimated that 3 to 5 terawatts (1 terawatt is 1012 watts) of baseload power generation can be practically harvested without affecting ocean temperatures or the environment of the planet. this is about the twice the global demand for electricity. Oceans are therefore an abundant renewable resource, offering a sustainable method for the production of electricity in the future. Most effectively, the technology can be used in equatorial zones with a temperature differential of at least 20°C (36°F) year-round.

Ocean Thermal Energy Conversion (OTEC) is a marine renewable energy technology that harnesses the solar energy absorbed by the oceans to generate electric power. Surface water is naturally warmer than deep ocean water due to the sun's heat, creating the ocean's thermal gradient. The Ocean Thermal Energy Converter uses the warm surface water of the ocean, which has a temperature around 25°C (77°F), to vaporize a low-boiling fluid, such as ammonia. A turbine coupled to a generator produces electricity when the vapor expands. When the vapor reaches the deeper ocean layer, it is cooled by seawater that is approximately 5°C (41°F) in temperature. The working fluid is then condensed back into a liquid so that it can be reused. It is a continuous cycle that generates electricity. Temperature differential plays a significant role in determining the cycle's efficiency. Efficiency increases with a larger temperature difference. Benefits

Working principle

An OTEC works by vaporizing a low-boiling point working fluid, such as ammonia, by utilizing warm ocean surface waters of around 25°C. Electricity is produced by expanding the vapor and spinning a turbine coupled to a generator. The vapour is then cooled by seawater that has been pumped from the deeper ocean layer, where the temperature is about 5°C. As a result, the working fluid can be regenerated into a liquid and reused. Electricity is generated continuously in this cycle.

Energy can be generated from the ocean in several different ways, including generating energy from waves, tides, and ocean temperatures. In OTEC systems, closed-cycle systems and open-cycle systems are most common. The working fluid, condenser, and evaporator are all necessary parts of both systems.

11.2 Closed Cycle OTEC

In a closed-cycle OTEC system the warm water is pumped over a closed loop pipe filled with a liquid with a low boiling point, likely ammonia as can be seen in Figure 11.1. The warm surface water evaporates the liquid into a vapor and this vapor turns a turbine. After the vapor passes through the turbine, cold deep seawater is pumped over the pipe which cools the vapor and condenses it back to a liquid. This liquid is then returned back to the beginning to begin the loop again and both the warm and cold seawaters are discharged back into the ocean.

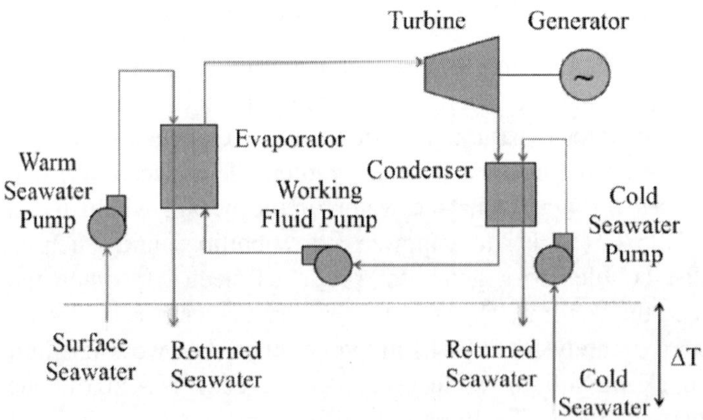

Fig. 11.1: Closed cycle OTEC

11.3 Open-Cycle OTEC

In an open-cycle OTEC system (Figure 11.2), the warm seawater is evaporated instead of the low boiling point fluid used in a closed-cycle system. During the process of pumping surface water into the low pressure chamber, the pressure in the chamber causes the seawater to evaporate and turn into water vapor, which in turn turns the turbine. As a result of the cold water condensing the water vapor back into water, the vapor is converted into water again. Compared to a closed-cycle system, an open-cycle system does not require dealing with low boiling point fluids, which may be hazardous if they escape into the environment. In addition, when evaporating warm sea water condensed back, it becomes fresh water, which could be shared with local communities in need. For the pressurized chamber to be able to evaporate the water, it has to have zero leaks, so maintenance is required.

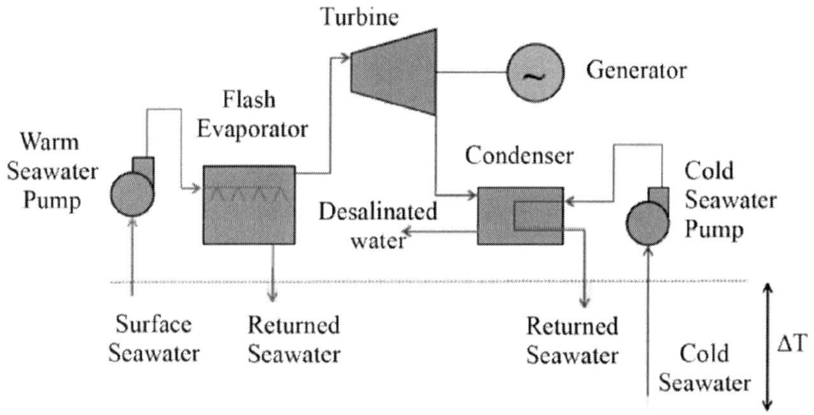

Fig. 11.2: Open cycle OTEC

11.4 Hybrid OTEC

A hybrid-cycle ocean thermal energy conversion system combines the best features of the two systems above by using both seawater and an ammonia-based working fluid which can be seen in Figure 11.3. In the hybrid OTEC system, warm seawater is vaporized into steam, similar to the open-cycle in a vacuum vessel. After mixing ammonia with warm seawater, the vapours are used to drive the turbine. The evaporated ammonia is then separated from the steam/water, re-condensed with the cold temperature, and reintroduced into the closed loop. By combining two systems in one, a hybrid system is more efficient.

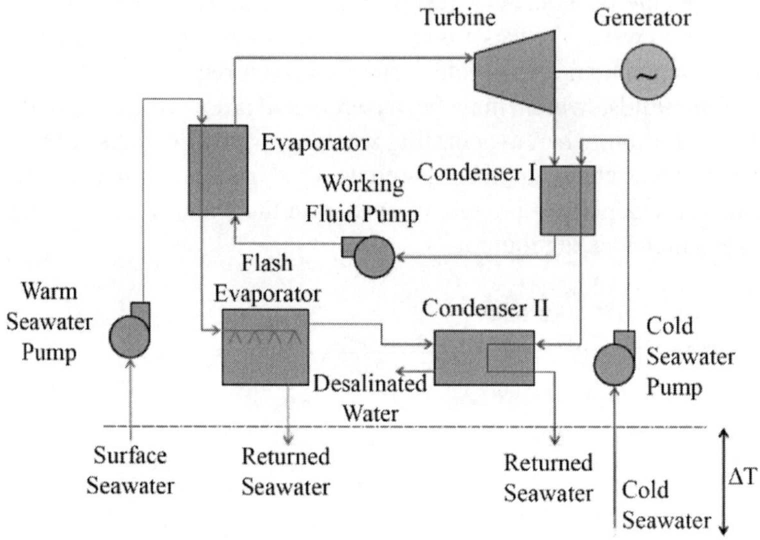

Fig. 11.3: Hybrid OTEC

Benefits of OTEC system

a) OTEC's distinctive feature is the ability to supply baseload electricity day and night (24/7), all year long.

b) Furthermore, OTEC provides the possibility of co-generating other synergistic products, such as fresh water, nutrient solutions to enhance fish farming, and seawater cooled greenhouses to enable food production in arid regions.

c) Cold water can also be used in building air-conditioning systems. Energy savings of up to 90% can be realized

Multiple Choice Questions

1. What type of working fluids do closed-cycle ocean thermal energy conversion (OTEC) systems use?

   a) Low boiling point fluids at atmospheric pressure

   b) High boiling point fluids at atmospheric pressure

   c) Low boiling point fluids at 100 bar

   d) High boiling point fluids at 100 bar

2. What is the warm surface seawater pumped through in a closed-cycle ocean thermal energy conversion (OTEC) system?

   a) Heat compressor　　b) Heat exchanger

   c) Fluid compressor　　d) Turbine

3. How is the vapourised working fluid condensed in a closed-cycle ocean thermal energy conversion system?

   a) Warm surface seawater　　b) Fresh warm water

   c) Cold water from deep ocean　　d) Fresh cold water

4. Which of the following best describes the open-cycle OTEC?

   a) Warm surface seawater → heat exchanger → working fluid → turbine

   b) Warm surface seawater → low-pressure container → turbine/generator

   c) Deep seawater → heat exchanger → working fluid → turbine

   d) Deep seawater → low-pressure container → turbine/generator

5. Why does an open-cycle OTEC use low-pressure container?

   a) To solidify the warm surface seawater

   b) To evaporate the deep seawater

   c) To boil and evaporate the warm surface seawater

   d) To solidify the warm surface seawater

6. Which of the following best describes the working of an ocean thermal energy conversion plant?

   a) Oceanic water → evaporator → turbine/generator → electricity

   b) Warm surface oceanic water → evaporator containing working fluid → turbine/generator → electricity

   c) Cold surface oceanic water → electricity →evaporator containing working fluid → turbine/generator

   d) Cold deep oceanic water → electricity → evaporator containing working fluid → turbine/generator

7. What is the byproduct of an ocean thermal energy conversion system?

   a) Electricity b) Clean water

   c) Water vapour d) Cold water

8. Open cycle ocean thermal energy conversion systems use _____ as the working fluid.

   a) Vapour from rivers b) Water from rivers

   c) Vapour from seawater d) Seawater

9. _______ turbine is used in closed cycle ocean thermal energy conversion.

   a) Horizontal b) Low-pressure

   c) High-pressure d) Vertical

10. Which of the following is used as working fluid in closed cycle oceanic thermal energy conversion systems?

    a) Thermohaline circulation b) Temperature gradient

    c) Greenhouse gases d) Refrigerants

11. Which of the following are types of systems used in ocean thermal energy conversion?

    a) Horizontal and vertical b) Vertical and open cycle

    c) Open cycle and closed cycle d) Horizontal and closed cycle

12. Why is ocean thermal energy conversion a renewable resource?

    a) Because the temperature gradient lasts for a short period of time

    b) Because the upwelling of cold water from the deep ocean is replaced by downwelling of surface waters

    c) Because ocean water is available in plenty

    d) Because of sun's heat

13. How does the cold and denser water masses sink to the depths of ocean?

    a) Thermohaline circulation b) Temperature gradient

    c) Density gradient d) Freshwater fluxes

Answer

1	a	2	b	3	c	4	b	5	c	6	b	7	d
8	c	9	b	10	d	11	c	12	b	13	a		

12

Magneto Hydrodynamic

12.1 Introduction

A magneto hydrodynamic energy conversion system or MHD system converts heat energy directly into electricity. An MHD generator produces electricity via the interaction of a moving fluid with a magnetic field and ionized gas. MHD conversion does not involve any mechanical conversion intermediary. As a result, this process offers substantial fuel savings because the process of producing mechanical energy and then converting it to electrical energy is eliminated. It is possible to generate electricity in large quantities using MHD power plants while reducing the impact on the environment. During his Bakerian lecture in 1832 to the Royal Society, Michael Faraday presented the concept of MHD power generation for the very first time. At Waterloo Bridge in Great Britain, he conducted an experiment to measure the magnetic field from the flow of the river Thames.

12.2 MHD Generator

MHD generators can be considered as a fluid dynamos. In the MHD generator, a conducting gas plasma replaces the metal conductor in the mechanical dynamo so as to create current in the conductor. In a magnetic field, a conductor moves perpendicular to the magnetic field and its movement causes an electrical field to be created. This is the principle behind conventional rotary electricity generators, discovered by Michael Faraday. The magnetic hydrodynamic generator (MHD) generates power directly by interacting with rapidly moving streams of fluid, usually ionized gases or plasma. A MHD device converts heat or kinetic energy into electrical energy. There are no moving parts in an MHD generator, so vibrations, noise, and wear and tear are limited to a minimum. MHDs have the highest thermodynamic efficiency since they operate at higher temperatures than mechanical turbines. Based on Fleming's Right Hand Rule, when conducting plasma moves through a magnetic field, a voltage (and associated current) is generated across the plasma, perpendicular to both the plasma flow and the magnetic field.

12.3 Principle of MHD Generation

Faraday's law of electromagnetic induction forms the basis for MHD power generation. Current flows across the terminals of a conductor when it is moved relative to a magnetic field, thereby causing voltage to be induced in the conductor. Magnetic and electric fields are used to affect the flow of a conducting fluid in a magneto hydrodynamic generator, as shown in the figure below. The conductor in a conventional generator or alternator consists of copper coils or strips, but in an MHD generator (Figure 12.1), hot ionized gas or conductor fluid replaces the solid conductor. The transverse magnetic field of a channel or duct circulates a pressurized and electrically conducting fluid. In order to transmit power to a load attached to the channel walls, electrodes are located at right angles to the magnetic field. MHD generators use electrodes instead of brushes as in conventional DC generators. MHD generators develop DC power and convert it to AC with the help of inverters. Due to the expansion nozzle, the gas pressure is reduced and the plasma speed increases (Bernoulli's Law) through the generator duct. In addition, the pressure drop causes the plasma temperature to fall (Gay-Lussac's Law), thereby increasing the plasma resistance, so there must be a compromise between Bernoulli and Gay-Lussac. A compressor is used to increase the combustion rate of the fuel using exhaust heat from the working fluid, but much of the heat is wasted unless it can be recycled.

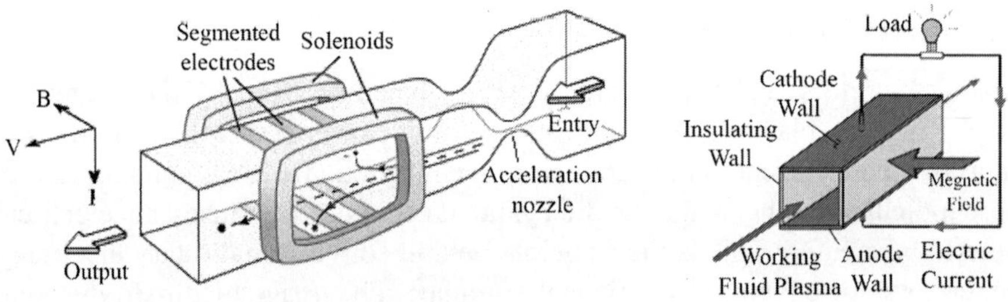

Fig. 12.1: Principle of MHD power generation

12.4 Components of the MHD

Gas Plasma

A high-conductivity plasma is critical to the MHD system since it is dependent on the plasma for its operation. The atoms or molecules in the plasma need to be ionised in order to achieve high conductivity. It is the flow of positively charged particles in the plasma that induces the current in a circuit enclosed outside the magnetic field. Gas can be ionized in different ways using adequate energy. A gas can be heated or irradiated using X-rays or Gamma rays to accomplish this. In order to enhance the ionization and conductivity of burning gas plasmas, small amounts of seed materials, such as potassium carbonate or cesium, are commonly added.

Containment

Since plasma temperatures are typically over 1000°C. Due to these extreme temperatures, ducts containing plasma must be made of non-conducting materials. Electrodes must of course be both conducting and heat-resistant.

12.5 The Faraday Current

Electromagnets provide the magnetic field through which plasma flows, and two electrodes placed perpendicular to it are used to generate the voltage across the plasma. Faraday current refers to the flow of current across the plasma between these electrodes. As a result, the MHD generator produces its primary electrical output.

12.6 Hall Effect Current

A Hall Effect current is generated when the high Faraday output current which flows through the plasma duct into the load reacts with the applied magnetic field. An energy loss caused by a current along the plasma axis. A total current is produced by adding the axial (Hall effect) and transverse (Faraday) current components. There will be an energy loss due to the Hall effect current unless it can be captured in some way.

12.7 Power Calculation of MHD

As per Fleming's Right Hand Rule, a voltage is generated across the conducting fluid when it flows through a magnetic field, which is perpendicular to both the fluid flow and the magnetic field. In a MHD generator, a conducting fluid passes through a magnetic field B according to Fleming's Right-Hand Rule. During a constant magnetic field, charged particles moving with a velocity v in a conducting fluid are governed by the Lorentz Force Law. The vector equation below gives the simplest description of this process.

$F = Q (v \times B)$

Where,

F = Force acting on the particle

Q = Charge of the particle

v = Velocity of the particle

B = magnetic field

The vector ‘F’ is perpendicular to both ‘v’ and ‘B’ according to the right-hand rule.

12.8 Types of MHD Cycles

The MHD cycles can be of two types, namely

1. Open Cycle MHD
2. Closed Cycle MHD

12.8.1 Open Cycle MHD System

MHD generators are generally used with conventional steam power plants (Figure 12.2). Fuels used include coal, oil, and natural gas. By compressing the surrounding air, the air compressor increases its pressure. A temperature of about 1100°C is then reached in the combustion chamber after the gases have been heated in the low-temperature and the high-temperature air preheaters. Fuel is burned in the presence of hot air in the combustion chamber. In order to increase their electrical conductivity, the hot gases are seeded with 1% alkaline potassium. MHD ducts receive these hot gases at 2300°C to 2700°C temperature as they expand within the nozzle to high velocity. In some cases, gas (plasma) can directly generate DC power by converting its internal energy. DC power is then converted into AC power by a DC-AC inverter. High temperature air preheaters use the hot gases left in the MHD duct to heat the air. As a result of these gases, steam is generated from feed water in a heat exchanger (steam generator). Steam is generated in a conventional steam power plant and used to generate electricity. Compressed air is heated by a low temperature air preheater using the remaining heat from gas. In order to recover hot gases, seed recovery units are used (electrostatic precipitators). In addition, these gases can be recycled again by compressing again, heating, and re-exhausting into the environment through a chimney.

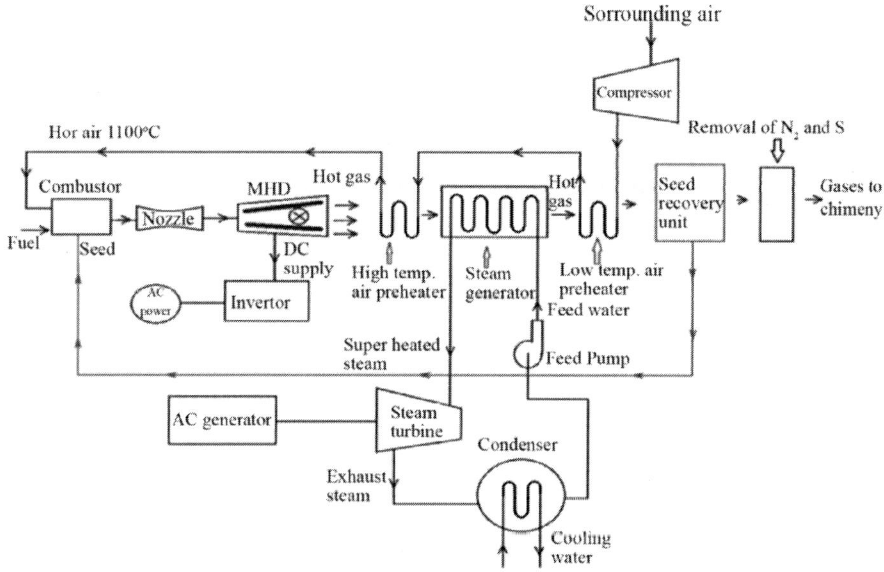

Fig. 12.2: Open cycle MHD system

12.8.2 Closed Cycle MHD System

In a closed-cycle MHD, the working fluid is circulated in a closed loop. The working fluid is therefore inert gases or liquid metals. Typically, liquid metals have good electrical conductivity, so they don't require a lot of heat from their combustion materials. A closed loop system has no inlets or outlets for atmospheric air, unlike an open loop system. Heat transfer is greatly simplified since the same fluid circulates repeatedly. A nuclear reactor heats inert gas helium seeded with cesium to a very high temperature. Helium gas is passed through nozzles to increase its velocity before it is supplied into MHD ducts to produce D. C. power which is converted into A. C. power by an inverter. A steam generator converts feed water into steam to be used in a conventional steam power plant by passing it through a heat exchanger called a steam generator. After being cooled in the cooler, the helium gas leaves the steam generator and is compressed in the compressor before being reinjected into the reactor. A gas turbine plant may be used instead of a conventional steam plant as the bottoming unit in the above system. MHD systems with seeded inert gas are the most promising.

Liquid metal MHD system

A closed cycle liquid metal MHD system that replaces helium gas with liquid metal (potassium) as shown in Figure 12.3. As liquid potassium passes through the nozzles of the nuclear reactor, it is heated to increase its velocity. A separator separates the vapour that is formed during the passage of the nozzle and pumps it back into the reactor. A high-speed MHD duct generates DC power by passing liquid metal through it. A heat exchanger (steam generator) heats liquid potassium to a relatively low temperature before it is pumped into the reactor. As a result, heat is transferred from the boiler to the feed water, which is converted into steam. As shown in Figure 12.3, this steam is used in conventional steam power plants.

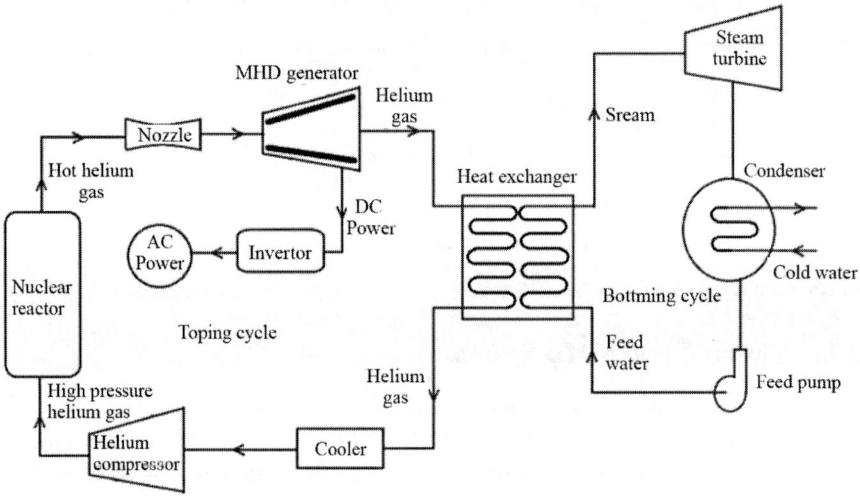

Fig. 12.3: A closed liquid MHD system

12.9 Advantages of MHD System

a) As compared to convection-type power plants, it converts heat energy directly into electricity.

b) Power produced by these systems is pollution-free and reliable.

c) In comparison to conventional power plants, small plants generate higher power.

d) Compared to conventional power plants, the capital cost is comparable.

e) It is cost-effective to operate and maintain.

f) A MHD plant can operate on any type of fuel, and its efficiency is 50%, compared to a conventional plant's maximum efficiency of 40%.

12.10 Disadvantages of MHD System

a. A high amount of fluid friction and heat transfer occurs in this system

b. To make gases electrically conductive, MHD generators operate at high temperatures. However, the electrodes must be relatively cooler

c. Due to the relatively colder temperature of gases around electrodes, it increases their resistance.

d. Voltage drops across gas films are large

e. Due to the need magnets, MHD systems are more expensive.

13

Energy Conservation in Agriculture

13.1 Introduction

At a national level, the agricultural sector is one of the major potential users of energy in the Indian context. Further, it is often highly dependent on the supply of electricity and fuel in order to function properly. The term "agriculture" refers to all on-farm activities required for farming, raising animals, and associated activities associated with farmers, growers, and entrepreneurs. Fuels such as petroleum and electricity are primarily used for the operation of pump sets and other operations. In the future, this may result in problems with supply and price. Moreover, India has relatively small reserves of crude oil and will eventually be forced to import more crude oil to make up for the lack of reserves. In order to save energy and money on the farm, farmers must identify opportunities for energy conservation on the farm in order to maximize energy efficiency. The cost and potential savings of each conservation opportunity should be evaluated independently using a standardized format. This will enable them to be compared with each other so as to determine their relative benefits.

There are three main categories of energy conservation opportunities that can be categorized as follows:

(1) Improvements in normal operations: These include simple tasks associated with housekeeping and maintenance, which can be done at relatively low or no cost to the farm. In other words, it indicates an improvement in existing tools and equipment that are used in various tasks.

(2) Changes to methods of operation: These can be implemented using resources available "off the shelf" such as equipment and technology. By replacing old and inefficient tools, more effective & efficient tools can be implemented.

(3) Change of system or process: Such a change involves both capital expenditure and redesigning of a production process that involves the changing of a complete process and tools of a different design and type.

The concept of energy conservation is far from implying a reduction in energy consumption, but rather implies the utilization of available energy in a more efficient manner. In a developing country such as India, the gap between supply and demand would result in an energy shortage. To reduce the gap between the growing demand for energy and its supply, energy conservation has become a necessity. Because our country has limited conventional energy sources and financial constraints to invest in capital-intensive energy sources, energy conservation studies serve the following purposes:

(1) Reducing the amount of investment required to meet the growing demand for energy is the first step.

(2) By adopting energy conservation techniques within the existing set up, we can reduce the gap between supply and demand within a short period of time.

(3) The third objective that needs to be achieved is to reduce the cost of energy made available to the farmers to meet the needs of different agricultural activities.

(4) The use of new and renewable energy sources as a means of replacement.

The following items could be included in a standard format for planning each energy conservation opportunity:

a) Identification of opportunities within the existing setup.

b) A description of current practice, i.e. the present pattern of energy consumption.

c) Observations regarding current energy consumption.

d) Recommendations for long-term planning.

e) It is expected that energy savings will be realized.

f) A study of the economic impact of energy conservation practices.

13.2 Farm Energy Management Approach

To begin managing energy on a farm, it is important to determine the total energy consumption, which can be easily calculated based on farm records. In this calculation, the total annual energy requirements and the associated costs are taken into account. The information would be as follows:

a. Pattern of energy consumption, energy costs, and production levels in recent years.

b) Details that might help in allocating energy used, at least approximately, to particular products or farm activities, such as 12% for grain drying, 10% for homesteading.

c) Calculation of the energy required per unit of production.

Using the above figures, one can evaluate the success of energy management programmes over the coming years based on the average energy intensity of farm products. Ultimately, the goal is to reduce energy intensity so that each unit of production consumes less energy.

After completing the analysis of accounts, the next phase is the farm energy audit. Its purpose is to:

a) Organize energy use according to function, location, or another appropriate classification.

b) Observe processes and facilities from an energy use perspective.

c) Identify opportunities for energy conservation.

During the audit, information is collected about the farm's major energy-consuming devices. Ideal information would be the operating schedule and rate of energy consumption for each device, but in practice, estimates may be required. Observing first-hand what is happening with respect to energy consumption is often useful for determining where measurements are necessary before a realistic assessment of potential changes can be made.

13.3 Energy Conservation Measures in the Agricultural Sector

Agriculture includes cropping industries, livestock and poultry industries, horticulture, and other industries with potential energy consumption. In this regard, it is worthwhile to implement different energy conservation measures in different areas.

13.3.1 Energy Saving in Farming Operations

In the cropping operation, a significant amount of energy is being consumed in tillage and seeding.

(A) Minimum Tillag: In agriculture, tillage refers to mechanically manipulating the soil surface in order to create a good tilt bed. In tillage operations, both commercial energy sources (petroleum) and animate energy sources (humans and animals) are used. The most significant energy savings can be achieved by reducing liquid fuel consumption and animate energy consumption by adopting minimum tillage methods. Depending on the region, adopting minimum tillage techniques can save a great deal of fuel.

The farming system does not need to be changed in order to achieve some energy savings. Choosing the right system, for instance, can result in a reduction in overall energy consumption.

a) Fallowing System: The savings are greatest on heavy soils where long-term fallows are repeatedly cultivated for weed control. On certain light soils (depending on the season and the level of need), paddocks are treated by conventional methods with reduced tillage. It is possible to minimize tillage of fallows by using chemicals such as glyphosate and paraquet/diquat mixtures. There is the potential for significant fuel and energy savings. However, herbicides used in some situations have a high indirect energy content.

b) Elimination of some overworked tasks. In the same areas in our context, soils are often overworked.

c) Direct drilling (the placement of seed and fertiliser into grooves cut simultaneously in uncultivated soil) and shallow cultivation can reduce cultivation requirements as well as soil erosion.

d) Choosing and using implements with low drafts.

e) Pre-emergent applications of trifluralin can reduce cultivation requirements.

f) The benefits of minimum tillage include not only energy conservation, but also soil conservation, time and money savings, extension of the growing season, improved tractor support during drilling and spraying, and timely sowing.

g) The tractor operation can also be changed to save energy.

In this context, some of the tillage operations can be described as follows.

Compared to smaller powered tractors, tractors with larger engines consume a lower amount of diesel per hectare of operation compared to tractors with smaller engines. There may be a need to balance this advantage against the need for a smaller tractor to perform daily farm operations or when smaller machinery is being utilized.

The weight of the tractor has a considerable impact on available power. Tractors that are too heavy or too light produce less power at the drawbar than tractors that are just the right weight.

Diesel engines may achieve maximum fuel economy when working under high load conditions. Therefore, it is important that implements are well matched. The engine can be throttled back to allow a more fuel-efficient loading to be achieved, or the gear can be changed to maintain a constant ground speed if this is not possible. In the event that the exhaust suddenly shows more black smoke than normal, the engine is overloaded. Unburned diesel fuel is being discharged into the exhaust gases by the engine, which is wasteful and expensive. The engine is unable to burn all the diesel fuel supplied through the injector.

In addition to engine speed, fuel economy is also influenced by engine speed. Figure 13.1 illustrates how correct speed affects fuel economy and engine output.

The majority of tractor engines are not designed to run continuously under full fuel or overload conditions. The governor control should be moved to the fully open position in order to determine whether or not the engine is overloaded.

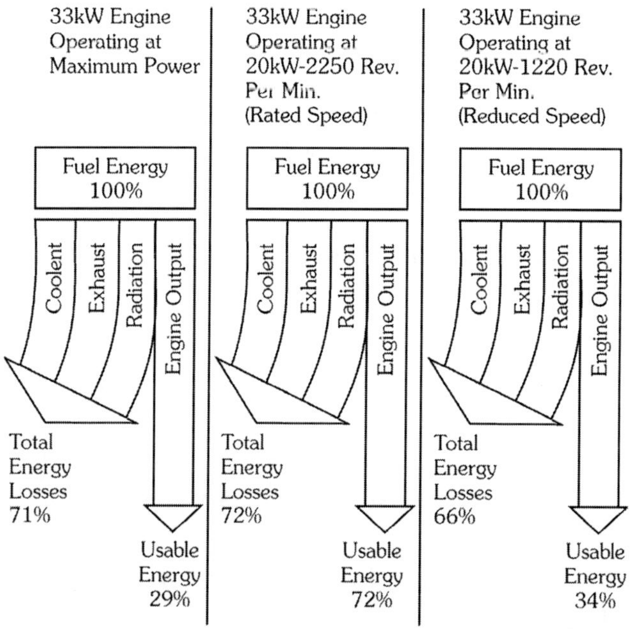

Fig. 13.1: Efficiency of Diesel Engine

An engine that starts up immediately does not have a problem with overloading, and can be dropped back to its original speed without any problems. If the engine is either unable to pick up speed or is very sluggish, it may be overloaded. Change down a gear and increase the engine governor setting in this situation.

There is a direct correlation between the level of wheelslip and the efficiency of converting engine power to drawbar power. Wheelslip that is too low or too high can have serious consequences. As a result of operating at 7 or 25%

wheelslip instead of 12%, approximately 14% more fuel would be required to accomplish the same task. Wheelslip levels must be close to optimum in order to achieve good fuel economy. Table 13.1 indicates the optimum levels of aiming for.

Table 13.1: Optimum Level of Wheel Ship

S.No.	Tractor type	Cultivated Soil	Firm Soil
1.	2 - WD tractor	7-11% slip	10-15% slip
2.	4 - WD tractor (Imported)	6-10% slip	8-12% slip

h) (h) Fertilizer application at seeding rather than as an extra pre-seeding operation results in an overall reduction in energy consumption. Practicality depends on whether fertiliser can be handled in bulk at seeding and how critical seeding timeliness is.

i) Proper engine maintenance is essential in order to achieve good fuel economy. The air filters, injectors, and fuel pump are the engine components that have the greatest impact on fuel economy. It is imperative that these three items work at peak efficiency in order to achieve good fuel economy. Items such as value adjustment, piston ring wear, fuel pump timing, and fuel system leaks and damages. It is possible that cooling fans may reduce engine performance and, as a result, fuel efficiency. The following symptoms may serve as warning signs to an engine that requires service:

- Emissions of smoke have been linked to worms, inefficient injectors, and/or clogged air cleaners.
- The presence of fumes from dipsticks and other access holes is due to excessive interior pressure caused by worn valves and damaged piston rings.
- Low compression and fast cranking over at start-up - worn valves or damaged piston rings.
- The engine cranks slowly due to dirty, carboned-up oil, the wrong oil grade (or a flat battery).

13.3.2 Energy Savings in Live Stocks and Poultry Industries

(A) Sheep and Beef Cattle: Fodder for animals is one of the most energy-intensive products we produce as it requires a great deal of energy to produce and conserve. There is a great deal of potential for saving money in this area. To produce conventional square bales of hay, farmers must perform five operations: moving, raking (once or twice), baling, and picking up. It is an energy-intensive process that requires up to five operations to produce conventional square bales. In addition to feeding, there is also a process that requires energy. Compared to the production of silage or big round bales, the process of windrowing pasture and leaving it in the paddock uses less energy than the process of producing silage.

As drought reserves such as hay, silage, or grain need to be held in case of drought, year to year seasonal requirements can be dealt with with the minimum amount of energy by:

a) Maintaining stock at conservation stocking rates in marginal agricultural areas so that pastures in the paddocks can sustain stock throughout the year.

b) Aligning the sheep production cycle to the pasture production cycle.

c) Running easy care sheep in order to minimize handling for purposes such as fly control.

d) Avoid oversupervising the stock, particularly at lambing when too much supervision can be detrimental.

e) In lieu of a utility vehicle or other large vehicle, motor bikes or other two-wheelers may be used for stock supervision.

f) In order to maximize the efficiency of road transportation, wool bales should be pressed or densified to the maximum weight allowed.

(B) Dairying : In the dairy industry, does the extent of possible savings depend on the layout and efficiency of each dairy? There is a significant amount of energy consumed in the process of heating the water and cooling the milk. These are the areas that are most likely to benefit from conservation measures, since they are the most vulnerable. The following are some possible options for the dairy industry in terms of reducing its energy consumption.

a) The milk is pre-cooled by using plate coolers before it is placed into the vat to prevent it from getting too hot. To ensure that refrigeration equipment operates efficiently, clean the cooling fans of the condenser, and provide adequate ventilation to circulate air over the condenser.

b) Heat can be generated by equipment or extracted from milk while it is being cooled in order to heat water. The water passing through the plate coolers is used for udder washing. Where there are no plate coolers, the water jacket around the vacuum pump exhaust heats the water.

c) Install solar water heaters to pre-heat water or convert manure into biogas for use as a fuel for heating. Besides providing clean combustible gas for cooking, biogas from animal waste also produces enriched manure for increased agricultural output.

d) The use of drain water or waste wash water in the mixing of dung for the production of biogas is recommended. It serves as a seeding material for the production of enriched gas.

e) Do not set the hot water thermostat higher than is necessary.

f) In large dairy operations where water is abundant, wash down the yards with high volumes of water under low pressure rather than low volumes under high pressure.

g) Build holding yards on a slope towards the dairy so that backing gates are moved forward by gravity rather than by an electric motor. It is also recommended for large cattle yards.

h) Insulate buildings that are either cooled or heated. The use of insulation can save you money. By minimizing heat loss, a tremendous amount of energy can be saved.

i) It is important to minimize the wastage of hot water by lagging pipes and repairing leaky faucets.

j) For outside lighting, it is recommended to use fluorescent, mercury vapor, or high pressure sodium vapour lamps rather than incandescent lights. Lighting through biogas plant running on dairy waste, can also be useful in conservation of energy in dairy.

k) In order to reduce the amount of water pumped from dams and boreholes, install rainwater tanks near farm buildings.

l) Attach the timer switch to the pump control.

m) The growth performance of calves can be as good when milk substitutes are prepared with cold water as when they are prepared with hot water.

n) Replace defective and old dairy equipment and machinery, which consume a substantial amount of energy.

o) In the areas of heating water, milk heating, refrigeration, and lighting, it is recommended that new and renewable energy sources be substituted for conventional energy sources.

p) In addition, the preparation of animal feed requires a significant amount of energy. There are currently built-in solar cookers available for preparing animal feed.

q) Currently, solar dish type concentrators can be used for pasteurizing milk. However, this method requires a substantial amount of energy.

There are a number of ways in which you can save energy in addition to coordinating fodder conservation and feeding out. It is possible that silage heaps may be located near areas where fodder is cut, for example. Fodder can be strategically placed around the farm so that it does not need to be moved for feeding. This will reduce or eliminate the need to move it. The installation of hay racks along haystacks makes it possible to drop hay directly into the racks without having to lift it up. It is possible to reduce the need for green chop harvesters and feed-out carts by using electric fencing on strip grazing paddocks.

(C) Poultry : It is necessary for poultry to be kept in confined conditions with artificial lighting and air temperatures that are controlled artificially. Even though the requirements for the production of laying hens and broilers differ, constructing or retrofitting any shed according to low energy design principles will reduce the necessity for energy intensive controls.

Generally, poultry sheds should maintain air temperatures at 20-28 °C, while allowing adequate ventilation so that fresh air is provided and dust and smells are minimised, particularly ammonia.

The use of artificial lighting cycles is also sometimes used in our country as a means of optimizing production. Depending on the type of bird, the lighting requirements range from 5 lux for pullets to 10 lux for laying hens, 20-30 lux for baby chickens, and 3 lux for growing broilers. Whenever possible, use fluorescent lights with reflectors of a minimum rating. The use of translucent roof panels can reduce the need for artificial lighting, however, in the summer, they may cause problems with heating. There are light-sensitive switches available that are capable of automatically adjusting lighting conditions over the course of the day.

Brooding consumes a significant amount of energy. It is recommended that brooders be placed in the center of a poultry shed in order to maintain brooding temperatures with minimal energy consumption. It is then possible to place a plastic tent or a solar canopy around the brooders in order to reduce heat loss. Make sure the temperature is maintained at an optimal level by using thermostats, or checking the temperature regularly. Raising the litter shed off the ground will allow air to circulate under it and assist in keeping the litter dry.

The direct fired hot air brooder is more efficient than most other systems; however, it should only be used in closed, insulated sheds. If the walls of the shed are not insulated, a double plastic curtain with an air gap of approximately 10 mm is recommended. Consider solar heated hot water brooding systems when purchasing new brooding equipment.

Feedstuffs for poultry are energy-intensive, and thus are susceptible to increases in the price of fossil fuels. Indirect energy savings will be achieved by implementing efficient feeding systems in order to reduce wastage. The majority of the feed intake by poultry is used to maintain their body temperatures. Therefore, ensuring that the temperature in poultry sheds is controlled efficiently will result in optimum feed consumption. The fouling and loss of feedstuff is often caused by rodents. As well as causing damage to insulation, curtains, pipes, and wiring, they can also cause a substantial financial loss. Using single-dose rodenticides is a convenient method of controlling rodent populations. In spite of this, cats can often be equally effective at controlling rats and mice, provided they do not harm poultry.

In addition to poultry litter and poultry droppings, other wastes are also suitable for the production of biogas. As a result of this fuel gas can be used for heating, lighting, and cooking as well as for the production of enriched manure for agricultural purposes. Solar water heaters, solar refrigerators, and solar electricity generators are quite promising for poultry use. Regular farm equipment inspections and maintenance will ensure maximum productivity with minimal energy consumption.

(D) Housing for livestock and poultry: There are three main categories of energy-efficient strategies for maintaining an optimal temperature in livestock shelters:

i.) Ensure that sheds are designed and constructed correctly so that adequate lighting, ventilation, and airflow are available.

ii.) Maintenance of buildings and energy-saving equipment for raising animals, preparing food, and cooking should be carried out on a regular basis.

iii.) Planning and coordination of livestock and energy requirements based on breed type and season.

(i) Building design: Building design can reduce the need for artificial heating, cooling, and lighting. Building orientation, window placement, ventilation, landscaping, and insulation reduce operating costs, and, except for insulation, are not necessary to increase the initial construction cost. Observing the daily fluctuations in the shed will provide an indication of whether additional controls are required. When the weather is favourable, adjustable shutters allow natural ventilation to maintain the temperature, and the shed can be shut during extremely hot or cold periods. Integrating solar architecture principles also enables the building to be oriented according to its needs. This not only reduces the energy bills associated with animal husbandry, but also provides comfort for the animals as well.

The placement of trees and shrubs around buildings can also contribute to the maintenance of an ideal temperature. The placement of deciduous trees on the northern and western sides of a building will allow the sun to fall on the building in winter, but not in summer.

Controllable thermostats and lighting levels enhance the flexibility of building use and therefore provide the opportunity to maintain optimal conditions while using the least amount of energy. In the case of domestic buildings, low energy designs are available for general purposes. The full benefits of low energy design can be gained by switching on energy-consuming equipment only when necessary.

(ii) Maintenance: Regular maintenance of equipment ensures that it operates at maximum efficiency and that no energy is wasted. It is possible to reduce current energy bills by as much as 30 percent by closing holes and cracks, covering windows with plastic if they are not required for ventilation, and cleaning and servicing light fittings, heaters, fans, and thermostats. Furthermore, in order to prevent air leaks in winter, it is a good idea to cover equipment when it is not in use. It is also a good idea to block off fan openings when it is not in use.

(iii) Planning: Assuring that heating, cooling, and lighting are coordinated according to the weather conditions and the production phases of livestock and poultry ensures the minimum use of energy.

13.3.3 Energy Savings in Horticulture

The costs of horticultural crops are generally lower than those of other agricultural products since they consume less fuel. However, there are ways to conserve both fuel and electricity that are used for cultivation, irrigation, fertilisation, spraying, pruning, and harvesting activities.

(A) Irrigation: Low pressure irrigation systems require less energy than high pressure irrigation systems, so they are more energy efficient. In order to achieve optimum water use and distribution and to reduce the cost of irrigation to cover a given area, compromises are necessary. In the past, irrigation was performed using a low pressure furrow reticulation system. Eventually, this was replaced by high-pressure overhead irrigation, which in turn was replaced by low-pressure, low-level applicators such as minijets, microsprinklers, and drippers. The irrigation management systems currently recommended to growers aim to use and distribute water efficiently through the installation of improved irrigation systems. In addition, they include tensiometers, evaporation data, and other monitoring tools. Irrigation water should be supplied based on the crop's needs, climatic conditions, growth stage, soil type, and quality. Irrigation techniques such as drip irrigation, sprinkler irrigation, and sheet irrigation must be integrated into horticulture growing areas.

There are many factors that must be considered when selecting the most effective and economical irrigation system. These factors include the characteristics of the site, the crop, the initial capital costs, the operating costs, labour requirements, and other equipment requirements. Energy consumption by irrigation systems is greatly affected by pump efficiency, which depends on how well the flow and head requirements match the pump's performance. Due to the rising cost of fuel, it is becoming increasingly important to select the type of energy used to operate irrigation pumps. As shown in Table 13.2, there are a variety of current options for powering irrigation pumps.

In order to save energy, dams should be carefully positioned around the farm. Water pumped from dams requires much less horsepower than water pumped from boreholes, although evaporation losses can be greater. Pumping bore water into dams can be accomplished using windmills or solar-powered systems, which operate intermittently when conditions are favorable. There are times when dams may be built up a slope so that adjacent areas can be irrigated either through flood irrigation or sprinkler irrigation by using gravity pressure from the dam. Modern irrigation devimini-sprinklers,isprinklers, main sprinklers, and drip irrigation are powered by solar photovoltaic deep well pumps.

(B) Fertilising : In spite of the fact that fertilizer is usually applied to crops at least once a year by a manure spreader, alternative methods of applying fertilizer can be found. These methods do not require separate operations. In high-tech horticulture plants, it is recommended to use a microprocessor-based fertilizer-irrigation system, which allows for the automatic application of fertilizer and the scheduling of irrigation in order to achieve optimal growth. Nutrients that are soluble in water can be applied via a spray unit, and if compatible with the spray mixture, can be applied on a routine basis. It is also possible to distribute soluble nutrients through irrigation systems, which is becoming increasingly popular. Spreading conventional fertilizer over unwetted soil is pointless when only a portion of the soil surface is wet, such as with droppers and microjets. Incorporating soluble nutrients into irrigation water always ensures that nutrients reach the roots of crop plants in wet oil.

The use of artificial fertilizers, such as nitrogenous fertilizers, is energy-intensive, and it is therefore recommended to minimize their use. It is important to use bio-fertilizer rather than artificial fertilizers.

Table 13.2: Energy Alternatives to Power Irrigation Pumps

S.No.	Energy Use	Initial Cost of Motor	Annual maintenance and operation cost as percentage of capital cost	Typical efficiency of fuel use	Engine life span (Yrs)	Disadvantage	Advantages
1.	Electricity	Low	5	85-90	25	If supply not available connection costly.	Lowest cost option if supply available, minimal maintenance
2.	Diesel	High	10	20-25	15	High initial capital cost Reliable for operation if well	loaded and maintained
3.	Petrol	Low	14	15-20	9	Limited range of motors	Good for low usage systems or standby units.
4.	LPG	Medium	12	15-20	15	Based on petrol engines therefore limited range Technology is not refined yet to commercialised	Secure fuel supply
5.	Solar photovoltaic cell	High	Low	-	over 20 years	Intermittent supply	No fuel cost, low maintenance Automatic Operation
6.	Biogas / Producer gas	Moderate	10	25-30	15	Require training to user	Environmental friendly operation

(C) Spraying : These are some of the energy saving measures recommended for spraying operations:

i.) Use concentrate sprays according to the needs of crops to reduce the volume of spray mixtures and, in turn, reduce the energy required to spread large quantities across the property.

ii.) The use of ultra-low volume (ULV) sprays could significantly reduce energy consumption. Additionally, this reduces the loss of quality during the application process.

iii.) Plan more efficient and perhaps less frequent spraying, by monitoring pest and disease populations with bait traps and weather data.

iv.) A dual- or multi-purpose application can be provided when compatible materials are combined in a spray mixture, thus reducing spraying costs.

(D) Frost Protection: Frost damage can be exacerbated by several practices, including the construction of windbreaks, the introduction of winter furrows and irrigation, planting of groundcover, the removal of overhead sprinklers, the planting of plants in hollows, and the lack of selection of frost-tolerant plants to use. There are several practices that exacerbate the damage caused by frost. The use of irrigation and the production of smoke around the perimeter of the land can also be used to protect the land from the effects of frost.

(E) Greenhouses: Greenhouses are also one of the most important requirements for the conduct of horticultural activities, which is why they play such a significant role in the field of horticulture. In order to reduce the amount of energy that is needed to maintain the conditions in a greenhouse, the following steps can be taken:

i.) Stokers should be adjusted based on the weather and the season.

ii.) Provide adequate attention and repair the openings in the greenhouse as soon as possible.

iii.) It is important to ensure that the temperature instruments are working properly.

iv.) The walls of the greenhouse would either be covered with plastic or a double pane of glass would be installed. It is always a good idea to use plastic that is UV-protected.

v.) The south wall of the house should be insulated during the winter months.

vi.) A greenhouse's surface area should be reduced as much as possible in order to reduce energy costs.

vii.) Make sure that the roof slope is facing north in order to take advantage of the sun and its other associated elements.

viii.) Investigate the use of solar heating systems that are active and can be used to provide heating for floors and roofs.

ix.) It is recommended to investigate the use of solar water or air heaters with heat storage and circulation systems. This will enable you to maintain the required temperature by using solar water or air heaters.

x.) Several renewable energy sources should be integrated into the energy conservation program in order to ensure that it achieves the best results. These sources include solar thermal energy and biogas.

13.4 Energy Conservation by Use of New and Renewable Sources of Energy

As part of the current energy quest, it has already been established that energy conservation can be accomplished by improving the efficiency of existing systems (i.e. minimizing heat losses) as well as by substituting these new and renewable sources of energy with those that are available in the local area and that are environmentally friendly. In the current state of technology, devices are available to harness these New and Renewable energy sources, such as solar, biomass, wind, and improved chulha.

13.4.1 Solar Energy

It is possible to harness solar energy in a decentralized manner. This reduces the costs associated with transmission and distribution, which account for more than half of the total cost of providing electricity. The most successful applications of solar energy are heating and cooling of buildings, distillation, drying, cooking, engines for pumping water, refrigeration, photovoltaic conversion, furnaces, and bioconversion. In our country, a hot water system, cooking system, and drying system are some of the most interesting rural applications.

There are many uses for hot water, including bathing, cleaning, laundry, as well as heating the space in which you live. To meet this need, there are a variety of units available (Hot plate collectors). A flat plate collector with a surface area of 2m^2 is required for the heating of 100 liters of water from room temperature to 60 °C at a cost of Rs. 15,000/- – Rs. 20,000/-. Depending on operating conditions, efficiency ranges from 30% to 40%. Hot water system performance is estimated as follows:

Heat energy required for heating 100 lit of water upto 60°C through conventional Source

Energy consumption per year = 365 x 4 = 1460 kWh.

One can conserve 1460 KWh of energy per year by replacing this system with an alternative source of energy (Solar energy). The capital cost of the solar flat plate collector recovered can be calculated on the basis of the pay-back period analysis, which results in a 4-year payback period.

The energy consumed for cooking is a major component of total energy consumption in rural areas at present. Solar energy offers a suitable solution by taking into consideration the importance of cooking and the availability of suitable energy sources. Solar energy is capable of cooking and keeping food warm. Food is cooked by raising its temperature to around 100 °C, which is the temperature at which most foods containing water are cooked. It takes approximately 2-2½ hours for a box-type solar cooker, which costs approximately 4500/- to cook food, depending upon the type of food and the time of year. In most cases, different items such as dal, rice, vegetables, etc., are cooked simultaneously in separate containers. At a time about 500 gm of rice can be cooked from this cooker.

Heat required for cooking 0.5 kg rice is calculated as follows:

$$\text{Sensible heat for rice} = 0.5\ \text{kg} \times 0.431\ \frac{\text{kcal}}{\text{kg}\ ^\circ\text{C}} \times (100 - 25)$$

$$= 18.27\ kcal$$

$$\text{Sensible heat for water} = 1\ \text{kg} \times 1\frac{\text{kcal}}{\text{kg}\ ^\circ C} \times (100 - 15)$$

$$= 85\ \text{kcal}$$

It is estimated that the total amount of heat required to cook 0.5 kg of rice is 103.27 kcal. This means that by using a single box type solar cooker we can save 37960 kcal of energy annually or 30% in the cost of cooking fuel. An average rural household consumes approximately 5 kg of fuelwood per day for cooking. If the solar cooker is used to supplement the cooking process, approximately 30% of fuelwood can be saved. Therefore, approximately 550 kg of fuelwood would be saved per household per year.

A solar drier uses flat plate collectors to heat materials and remove moisture. Typical driers have 1 m width by 2 m length with adjustable inclination for removing 1.3 kg moisture from the material per hour. The development of bigger units and the use of modular construction can be used for larger capacities. The drier›s performance can be estimated as follows:

$$Radiation\,absorbed\,by\,material = 800\frac{kcal}{hm^2}\times 2m^2 \times \frac{50}{100}$$

Where typical value of global radiation $800\frac{kcal}{hm^2}$ overall thermal efficiency 50%, and drying area 2m^2 is used. Latent heat of vapourisation of water at 30 ^{0}C is 540 kcal/kg.

Moisture removal rate = 800/540 = 1.48 kg/h.

Conventionally for removing 'moisture at a rate of 1.48 kg/h the heat requirement is calculated as follows,

$$\text{Sensible heat of water} = 1.48 \text{ kg} \times 1 \frac{\text{kcal}}{\text{kg °C}} \times (100\text{-}25)\ \text{°C} = 106.56 \text{ kcal}$$

$$\text{Latent heat of water} = 1.48\text{kg} \times 540 \frac{\text{kcal}}{\text{kg}} = 799.20 \text{ kcal}$$

Total heat required per hour = 905.76 kcal.

Therefore, it takes about 1 kWh of electricity to remove moisture at a rate of 1.48 kg per hour. A solar drier with an area of 2 m^2 can conserve one unit of electricity per hour for drying any food grain.

13.4.2 Biogas

The generation of biogas is a simple and inexpensive method for decentralized energy production under a variety of climatic and agricultural conditions. Anaerobic digestion of organic materials results in the production of methane and carbon dioxide. The digested slurry contains the majority of the nutrients - both in terms of NPK - and no fertilizer value is lost. It is therefore possible that biogas can serve the dual purpose of meeting some of the essential energy needs of rural communities. In addition, it can increase agricultural production at the same time.

Cooking, boiling water, and steam generation can be accomplished with biogas. Biogas can easily be used to fuel many oil-fired boilers, resulting in a reduction in oil consumption. In agriculture and village industries, biogas can be utilized as a fuel source. The fuel can be used to run diesel and petrol engines, which in turn can pump water. Electricity can even be generated from it, and the cost is low, so it is more acceptable.

It is estimated that only 10% of gobar is consumed in the production of gas. However, the amount of heat that can be obtained from burning the full amount of dung is greater than the amount obtained from burning A major reason for this is the very high thermal efficiency of gas (60%) compared to the poor thermal efficiency of dung cakes (11%). In the same way, the amount of manure obtained by the gas plant is 43% greater than that obtained by the pit. It has been observed that about 0.227 cum of gas is required for cooking, 0.127 cum of gas per lamp of 100 candle power for lighting and 0.5 cum of gas per horsepower for running a diesel engine. As a matter of fact, one cum of biogas can replace 1.62 liters of kerosene or 5 -8 kg of wood. Only a small amount of petrol is consumed during the starting process of a biogas engine compared to a petrol engine. The replacement of diesel fuel by dual-fuel engines could be as high as 80% for diesel engines modified to dual-fuel engines. Diesel engines consume approximately 0.5 m^3 of biogas per horsepower or 0.65 m^3 per kilowatt-hour. An irrigation pump, flour mill, chaff cutter, thresher, etc., can be powered by a dual fuel engine.

A community biogas plant makes biogas technology accessible to all sectors of the rural population, resulting in greater efficiency for the system. Moreover, it has been noted that a community biogas plant could be useful to farmers without their own dung for gas production. However, collectively on a payment basis, they might be able to take advantage of biogas for cooking & other uses. A community biogas plant with a capacity of 85 cubic meters can meet the cooking energy requirements of 40 average-sized families.For this CBP to be economically viable, biogas can be used for transportation, lighting, and cooking.. An annual expenditure consists of depreciation, interest on capital investments, maintenance costs, and cattle dung used in digestion as fertilizer. The income generated by the plant is derived from the sale of manure and the production of biogas each year.

13.4.3 Biomass

Direct combustion of biomass in chulha can be used as a fuel without any intermediate conversion. Firewood stoves are used by 90% of rural households, 75% of semi-urban households, and 25% of urban households in India. This results in a significant loss of forest every year. Traditional Indian household chulhas have a thermal efficiency of approximately 5 to 10%. There are a number of improved chulhas available that have thermal efficiency above 20% in order to maximize the utilization of firewood. Furthermore, the improved chulha saves fuelwood directly, as well as having a number of other socioeconomic and health benefits. A typical improved chulha saves about 915 kg of fuelwood per year.

It is possible to convert biomass into gaseous fuel for use in engines as a supplement or replacement for petroleum. It is not only an economically feasible proposition, but also a feasible one. Even biomass can be converted into liquid fuel for power production, reducing the need for conventional fuel.

13.4.4 Wind Energy

Wind energy can be very effectively harnessed for water pumping in coastal and hilly areas, desert areas, and other areas with strong winds. A typical wind mill starts lifting water at 12 kmph. It yields about 30 to 35 cubic meters of water per day, for approximately 75% of the days in a year, at a reasonably windy location. Installing wind pumps of a suitable size for irrigation and drinking purposes can reduce the amount of energy consumed in raising water for these purposes.

13.5 Some Tips for Energy Conservation

There are two major fields that can be classified as energy conservation in rural areas, which utilize conventional sources of energy in place of alternative energy sources. First of all, at former's houses, and then at various electrical apparatus and machines that were used in rural industries and agricultural operations. These included devices meant for rural development and other devices used in rural areas. The following are some of the basic tips that can be used in both of these fields in order to conserve energy:

13.5.1: Energy Conservation Tips for Farmer's Building

S.No.	Item	Why Important	How to Conserve
1.	Lighting (For domestic use)	For illumination purpose	Turn off lights when not in use.
			Install lighting only to that required for task.
			Consider replacing lighting with more efficient lamps incandescent to fluorescent to gaseous discharge lamp
			Consider using fittings which are more effective in directing the light to the work area.
			Consider integration of daylight with electric lighting.
			Ducted fittings will improve the operating efficiency of suitable lamps.
			Lamps and fittings should be cleaned as part to a regular maintenance programme. Utilize day lighting as practicable. Heat reflective coatings may make windows more useful.
			Control window brightness. If windows admit excessive light, the general lighting standard may have to be raised to reduce contrasts.

S.No.	Item	Why Important	How to Conserve
2.	Space Air-Conditioning and Heating Systems (Meant for Storage of Agriculture Produce)	The energy required by the heating, ventilating and air-conditioning system is used for fan motors, refrigeration systems and heating systems.	A complete analysis is required to minimize the energy required.
			Control settings should not be changed without considering the effect on the whole system:
			Clean filters on a regular basis.
			Lock out refrigeration system by a return air thermostat.
3.	Steam/Hot Water (Domestic as well as for Cottage Industries use)	Insulate lines.	Repair leaks.
			Insulate lines
			Check operation of traps.
			Use of insulated storage tanks to receive heated water during off-peak hours can improve operating efficiencies.
4.	Hot Water For Personnel Use	Heat is required to raise the temperature of the water from the entering temperature to the desired temperature.	Repair leaks at hot water taps.
			Check to see that hot water taps are shut off when not in use.
		Heating rate required is approxmately Power (k1N)= 4.2 x Flow Rate (1/5) x Temperature Rise (°C)	Reduce hot water thermostat setting to the lowest temperature which is acceptable.
			Separate system from space heating or other hot water system.
			Clean heaters regularly, following the manufacturer's recommendations. Should be flushed and- cleaned periodically.
			Check the standard of insulation on all hot pipework.
5.	Ventilation for domestic living and for Storage	Ventilation air be required for dilution and makeup.	Requirements for ventilation may have changed. Evaluate present requirements.
			Measure present ventilation. Use minimum possible but not below requirements of code.
6.	Openings in Structure (e.g., loading docks)	For easy and proper convenience in operation	Excessive infiltration of outside air requires extra energy to provide comfort.
			Close all unnecessary openings-unused exhaust fans, broken windows, structural openings.

S.No.	Item	Why Important	How to Conserve
			Enclose the trucks or railroad cars at the loading platforms.
			Could clear plastic strip partitions be of use?
			Investigate the possible use of automatic doors or air curtains.
7.	Wall and Roof Insulation	Most uninsulated walls and roof structures may allow 2 to 3 times the heat loss of an insulated structure.	Insulate roof and walls which reduce heat loss through the building structure.
8.	Windows and Doors	Heat loss by transmission/ infiltration.	Add weather stripping or seal windows
		Doors and windows on opposite walls permit excessive infiltration by direct flow of air through the building	Erect weather screens externally or internally to help protect doorways.
			Add door closers
			Lock, seal or change doors and/or windows on one wal
			Use door closers
			Break the flow of air by use of air locks and internal doors

13.5.2: Energy Conservation Tips for Rural Implements and Machines

S.No.	Item	Why Important	How to Conserve
1.	Furnaces used for Cottage Industries	Heat loss through furnace structure increases energy consumption.	Repair furnace linings.
			Inspect insulation for periodic maintenance and/or update.
			Check furnace doors and other openings for tighter closure.
			Reduce the holding temperature during idle time.
			Schedule work to reduce heat up and cool down periods.
			Periodically check on the furnace pressure control.
			If the furnace is being cycled, as in heat treating, investigate low thermal mass insulation.
2.	Burners for Domestic as well as Rural Industries	Either incomplete combustion or too great an amount of excess air wastes significant amounts of fuel.	Perform periodic combustion analysis and adjust burners.
			Maintain, repair or replace inefficient burners.

S.No.	Item	Why Important	How to Conserve
3.	Combustion Gases-Heat Recovery (For Rural Industries)	Products of combustion which leave at temperature higher than supply air represent a loss of energy.	Use exhaust gases to preheat combustion air by use of a recuperator heat exchanger.
			Use exhaust gases to generate steam or hot water and to heat secondary fluids or air for other uses.
4.	Process Heat Distribution System (Steam, Condensate, Hot Water and Chilled Water)	Any losses in the distribution system require greater energy input for the same end use.	Repair leaks in any line. Shut off or remove unused line.
			Provide for periodic maintenance checks on controls, valves, and accessories.
			Meter consumption and compare at regular intervals to identify unusual changes.
5.	Process Fuel Distribution System (Fuel oil, LPG, diesel, etc.)	Any losses in the distribution system require greater energy, input for the same end use.	Repair leaks in any line. Shut off or remove unused line.
			Provide for periodic maintenance checks on controls, valves, and accessories.
			Meter consumption and compare at regular intervals to identify unusual changes.
6.	Incineration of waste	Possible source of energy. Recovered heat can be used for process.	Investigate the possibility of incorporating a heat recovery system into incineration process.
7.	Cooling towers/ Evaporative Coolers (Essential device for Decentralised Power Generation Plant)	Heat rejection is required by many industrial processes such as die casting machines, plastic moulders and air compressors.	Investigate the possibility of using the heat for preheating boiler feed water or for hot water use, for heating stockpiles or other such low temperature application.
8.	Exhaust over Vats, Tanks, Grinding and Other Operations	Energy is required to heat the makeup air from the outside Energy is also needed to maintain the temperature of the bath.	Consider the use of heat recovery equipment from the exhaust air to the makeup air.
			Check the possibility of introducing outside air directly to operation without heating.
			Install covers over unsuitable for vats and tanks, consider insulating balls floated on the surface.

S.No.	Item	Why Important	How to Conserve
9.	Compressed Air System (For Rural Industries)	Energy is used to drive the air compressor.	Use coolest intake air available possibly from outside of building.
			Drive may be either electric motor, engine or turbine drive.
			Repair all leaks.
			Operate at the lowest air pressure required.
			Consider the possibility of using the heat from the after cooler for supplementing the plant space heat.
			Do not use compressed air for cooling equipment or for personal comfort. It is ventilation. Use air blowers expensive at required locations instead.
			Investigate the possibility of shutting down the compressor when not in use.
10.	Electric Power System (For Domestic as well as Agricultural Operations)	Transmission losses include transformer and fine losses. High peak demand and low power factor both increase the cost. They also increase the electric generating plant size required to service a given job.	De-energize transformers whenever possible.
			Clean transformer heat exchanger surfaces
			Investigate the possibility for scheduling use of power and for reducing demand.
			If power factor is low, additional load may involve expensive cabling costs; have a specialist recommend changes.
11.	Electric Motors (For Agricultural Operations)		Turn off equipment when no in use
			If operating efficiency is below 70%, investigate further
			If due for retirement, investigate higher efficiency motors and matching of motor to be load for best economic return.
12.	Controls -Systems and Settings	Controls determine the actual operating characteristics of the system.	Original system design and operation may have been based on different parameters and energy costs.
			A new look may "tune up" system for lower energy costs.

S.No.	Item	Why Important	How to Conserve
13.	Central System Boilers (For Rural Industries)	Combustion of fuels to generate steam or hot water.	Perform periodic combustion analyses and adjust burners.
			Consider heat recovery from flue gases.
			Clean, blowdown to maintain operating efficiency
			Could a small local electric boiler supply isolated low eve. requirements?
14.	Unit Heaters- Direct Fired (For Rural Industries)	Combustion of fuels for heating air.	Readjust burners for maximum efficiency.
			Clean heat exchanger surfaces
			Cycle fan operation.

Questions

1. Explain the energy conservation measure in the agricultural farm.
2. How energy can be saved in horticultural activities.
3. Classify the energy conservation opportunies.

References

Angrist SW. Direct Energy Conversion. 2nd ed. (Boston, AA: Allyn and Bacon, Inc. 1971).

Backus CE. Ed. Solar Cells, Institute of Electrical and Electronics Engineers (IEEE) Press, New York, 1976).

Ball RS. Natural sources of power. Constable, London, 1908.

Batten MI. English Windmills, Vol. I. Society for the protection of Ancient Buildings, Architectural press, London, 1930. Vol. II by Smith, D., 1932.

Beckman WA, et al. Design Considerations for a 50 Watt Photovoltaic Power System Using Concentrated Solar Energy", Solar Energy 1966; 10(3): 132-136.

Beedell S. Windmills. David & Charles, Newton Abbot, 1975.

Benko I. Possibilities of utilizing renewable energy sources. Procidia polytechnic, Mech Engg. 1981; 25; 67-86

Bradley WJ. Designing and Sitting Solar Power Plants," Consulting Eng. 1977; 48(3): 8084.

British Wind Energy Association, Wind Energy for the Eighties. peter perigrinus, Stevenage, 1982.

Bruno R. A correlation procedure for separating direct and diffuse insolation on a horizontal surface. Solar Energy 1978; 20: 97100.

Buch F, Fahrenbruch AL, Bube RH. Photovoltaic properties of five II□VI heterojunctions. Journal of Applied Physics, 1977; 48(4): 1596-1602.

Buckiust RO, King R. Diffuse solar radiation on a horizontal surface for a clear sky. Solar Energy 1978; 21: 503509.

Buckland S. Lee's patent Windmill, 17441747. Wind & Watermill Section, Society for protection of Ancient Buildings, London, 1987.

Calvert NG. Wind power principles, Their Application on the Small Scale. C. Griffin & Co. Ltd., London, 1979.

de Little RJ. The windmill Yesterday and Today. J. Baker London, 1972.

Demirbas A. Recent advances in biomass conversion technologies. Energy Educational Science and Technology 2000; 6: 19–40.

Dermatis SN, Faust, JW. Jr. Semiconductor Sheets for the Manufacture of Semiconductor Devices. IEEE Transactions Communications Electronics (New York: IEEE Press, 1963).

Devi BA, Kamalakkannan N, Prince PS. Supplementation of fenugreek leaves to diabetic rats. Effect on carbohydrate metabolic enzymes in diabetic liver and kidney. Phytotherapy Research 2003; 17: 1231-1233.

Duffie JA, Beckman WA. Solar energy thermal processes. Willey Interscience publication, 1974.

Duffie JA, Beckman WA. Solar Engineering of Thermal Processes. Wiley, New York, 1991.

Duffie JA, Beckman WA. Solar Radiation. Solar Engineering of Thermal Process, John Wiley and Sons, New York 1980:3-42.

Hammond AL. Photovolatics: The Semiconductor Revolution Comes to Solar. Science. 1977; 197(4302): 445447.

Mathur AN, Rathore NS, New and Renewable Energy Sources. Bohra Ganesh Publications, Udaipur 1996.

Panwar NL, Kothari R, Tyagi VV. Thermo chemical conversion of biomass - Eco friendly energy routes. Renewable and Sustainable Energy Reviews 2012; 16 : 1801-1816

Panwar NL, Rathore NS. Potential of surplus biomass gasifier based power generation: A case study of an Indian state Rajasthan. Mitigation and Adaptation Strategies for Global Change 2009; 14:711-720.

Panwar NL, Shrirame HY, Rathore NS, Sudhakar J, Kurchania AK. Performance evaluation of a diesel engine fueled with methyl ester of castor seed oil. Applied Thermal Engineering 2010; 30(2-3): 245-249.

Panwar NL. Energetic and Exergetic Analysis of Walk-in Type Solar Tunnel Dryer for Kasuri Methi (Fenugreek) Leaves Drying. Int. Journal of Exergy 2014; 4: 519-531.

Panwar NL. Thermal modeling, energy and exergy analysis of animal feed solar cooker. Journal of Renewable Sustainable Energy 5, 043105.

Rathore NS, Panwar NL, Kothari S. Biomass production and Utilization Technology Himanshu Publication, Udaipur (Rajasthan) 2007.

Rathore NS, Panwar NL. (2007). Renewable energy sources for sustainable development. New India Publishing, New Delhi, India. 2007

Rathore NS, Panwar NL. Design and development of energy efficient solar tunnel dryer for industrial drying. Clean Techn Environ Policy 2011; 13: 125-132.

Rathore NS, Panwar NL. Experimental studies on hemi cylindrical walk-in type solar tunnel dryer for grape drying. Applied Energy 2010;87: 2764-2764.

Rathore NS, Panwar NL. Experimental studies on hemi cylindrical walk in type solar tunnel dryer for grape drying. Applied Energy 2010; 87: 2764-2767.

Ravindranath NH, Hall DO. Biomass, energy, and environment: a developing country perspective from India. Oxford, United Kingdom: Oxford University Press; 1995.

Van Wylen GJ, Sonntag RE, Borgnakke C. Fundamentals of Classical Thermodynamics, John Wiley and Sons; 1994.

Wolff AB. The Windmill as Prime Mover, 2nd edn. Wiley, New York, 1885.

Wood BD. Solar energy measuring instruments. Solar Energy Engineering, Academic Press, London, 1977.

Index

A

Absorption 19
Acetogenesis 122
Acidogenesis 122
Air mass 18
Albedo 20
Alkaline fuel cell 141
Ammonia 162
Anaerobic digestion 121, 124
Anemometer 62
Aperture 27

B

Baler machine 108
Baler Machines 108
Battery 37
Betz's law 55
Betz's Limit 55
Biodiesel 111
Biogas 121, 189
Biogas production mechanism 122
Biogas technology 121
Biomass 77, 190
Biomass combustion power plant 80
Biomass energy 77
Biomass gasification system 85
Biomass pyrolysis 94
Blades 62
Brake 62

C

Central receiver systems 33
CIGS 40
Closed cycle MHD 169
Closed cycle OTEC 160
Coal 4
Cogeneration 98
Commercial energy sources 4
Concentrating solar collectors 26
Concentrating solar collectors 29
Concentration ratio 27
Containment 167
Controller 62
Conventional energy sources 3
Copper-Indium Gallium Diselenide 40
Counter flow gasifier 88
Crossflow gasifier 90
Cut out speed 56
Cut-in speed 56

D

Darrieus wind turbine 65
Deenbandhu model biogas plant 126
Digester 124
Disadvantage of PV system 39
Downflow gasifiers 89
Dye-Sensitized solar cells 40

E

Efficiency of solar cell 43
Electricity 13
Electrolysis 138
Electromagnetic radiation 15
Energy balance 25
Energy conservation 173
Energy sources 3
Energy 1, 9
Enthalpy 30

F

Faraday current 167
Farm energy management 174
Fertilising 184
Fixed dome biogas plant 126
Flat plate collector performance 25
Flat plate collectors 22
Floating drum plants 125
Fluidized bed boilers 81
Fluidized bed gasifier 91
Fossil fuels 4
Frost protection 186

Fuel cell 137
Fuel cell efficiency 147

G

Gas holder 124
Gas plasma 166
Gear box 62
Generator 63
Global warming 8
Global winds 52
Graphene solar cells 40
Grate stokers 80
Green houses 186
Greenhouse effect 7
Grid interactive System 44

H

Hall Effect current 167
HAWT 60
HAWT advantages 61
HAWT disadvantage 61
High speed shaft 63
Horizontal-axis turbines 59
Hybrid OTEC 161
Hybrid solar power system 45
Hydroelectric power 5
Hydrogen 137
Hydrolysis 122

I

Impulse turbines 153
INDEX
Inverter 49
Irrigation 183
I-V characteristic 40

J

Janta model biogas plant 127

K

KVIC biogas plant 125

L

Landfills 130
Liquid metal MHD system 169
Local winds 51
Low-speed shaft 63

M

Magneto hydrodynamic 165
Mesophilic digestion 123
Methanogenesis 122
MHD power generation 166
Micro hydropower 149
Microbial fuel cell 145
Minimum tillage 174
Molten carbonate fuel cell 143

N

Nacelle 63
Natural gas 5
Nature of wind 51
Non-conventional energy sources 3
Nuclear energy 5

O

Ocean thermal Energy conversion 159
Off-grid solar power system 44
Oil 5
On-grid solar power system 44
Open cycle MHD 168
Open-cycle OTEC 160
Organic photovoltaics 40
Organic urban waste 130

P

Parabolic trough collector systems 32
Parallel flow gasifier 89
Pelleting 103
Pelton wheel 153
Perovskite photovoltaics 40
Phosphoric acid fuel cell 141
Pitch 63
Poultry 182
Power coefficient 55
Power in wind 54
Power tower 33
Proton exchange membrane fuel cell 142
Psychrophilic digestion 123
PV system 39

R

Reaction turbines 154
Renewable energy 2, 6
Rotor 63

S

Savonius wind turbine 66
Series and parallel connection of solar cell 42
Sewage sludge 129
Silicon 39
Small hydro project 149
Solar arrays 38
Solar cells 37
Solar Chimney 35
Solar constant 18
Solar Energy 15, 187
Solar energy technologies 20
Solar modules 38
Solar panels 38
Solar photovoltaic 36
Solar PV plant 44
Solar PV system sizing 46
Solar radiation 15
Solar thermal application 21
Solar thermal power systems 31
Solid oxide fuel cell 144
Solid polymer fuel cell 143
Spraying 186
Standalone system 44
Steam reforming 137
Sun 15

T

Thermal efficiency 26
Thermolysis 138
Thermophilic digestion 123
Thin-film photovoltaics 39
Tower 63
Transesterification 112
Transport fluid 23
Turgo impulse wheel 154
Types of biomass pyrolysis 95
Types of gasifiers 87
Types of MHD Cycles 168
Types of solar cell 39
Types of solar collectors 21
Types of wind turbines 59

U

Underfeed stokers 80
Updraft gasifier 88

V

VAWT advantages 64
VAWT disadvantage 65
Vertical axis wind turbines (VAWT) 63
Vertical-axis turbines 60
Viscosity 30

W

Wind dirction 63
Wind energy 51, 191
Wind farm 67
Wind farm site selection 69
Working principle of PV 37

Y

Yaw drive 63
Yaw motor 63